Rapid Nondestructive Testing Technology-Based Biosensors for Food Analysis

Rapid Nondestructive Testing Technology-Based Biosensors for Food Analysis

Guest Editors

Yonghuan Yun
Jiangbo Li

MDPI

Basel • Beijing • Wuhan • Barcelona • Belgrade • Novi Sad • Cluj • Manchester

Guest Editors
Yonghuan Yun
School of Food Science
and Engineering
Hainan University
Hainan
China

Jiangbo Li
Intelligent Equipment
Research Center
Beijing Academy of Agriculture
and Forestry Sciences
Beijing
China

Editorial Office
MDPI AG
Grosspeteranlage 5
4052 Basel, Switzerland

This is a reprint of the Special Issue, published open access by the journal *Biosensors* (ISSN 2079-6374), freely accessible at: www.mdpi.com/journal/biosensors/special_issues/rapid_nondestructive_testing_technology_biosensors_food_analysis.

For citation purposes, cite each article independently as indicated on the article page online and using the guide below:

Lastname, A.A.; Lastname, B.B. Article Title. *Journal Name* **Year**, *Volume Number*, Page Range.

ISBN 978-3-7258-2826-5 (Hbk)
ISBN 978-3-7258-2825-8 (PDF)
https://doi.org/10.3390/books978-3-7258-2825-8

Contents

biosensors

Editorial

Rapid Nondestructive Testing Technology-Based Biosensors for Food Analysis

Yong-Huan Yun [1,*] and Jiangbo Li [2,*]

1 Key Laboratory of Tropical Fruits and Vegetables Quality and Safety for State Market Regulation, School of Food Science and Engineering, Hainan University, Haikou 570228, China
2 Intelligent Equipment Research Center, Beijing Academy of Agriculture and Forestry Sciences, Beijing 100097, China
* Correspondence: yunyonghuan@hainanu.edu.cn (Y.-H.Y.); lijb@nercita.org.cn (J.L.)

Food analysis plays a vital role in ensuring the safety and quality of food products. Traditionally, food analysis is performed through destructive testing, which involves the physical or chemical alteration of the sample. In recent years, there has been a significant increase in the demand for food safety, quality, and authenticity. To meet this demand, rapid nondestructive testing (RNT) technology has become an important tool in the food industry. It refers to techniques that can analyze the properties of a sample with the advantages of real-time and in-situ analysis, without causing any damage or alteration to the sample.

Biosensors are analytical devices that employ chemical and biological components to detect and quantify the existence or concentration of specific substances in a given sample. They have diverse applications throughout the stages of food production, processing, and storage, enabling quality control, contamination detection, and hazard identification. Over the past few years, significant progress has been made in the development of biosensors in RNT technology for food analysis. Researchers have explored various detection methods, such as electrochemical, optical, and immunological, to develop biosensors that can detect specific food contaminants, including pathogens, pesticides, and mycotoxins, with high precision and sensitivity. Moreover, biosensors have also been developed to detect food quality parameters, such as nutritional components, freshness, ripeness, spoilage, and authenticity. These biosensors can help food producers and processors monitor the quality of their products, reduce waste, and increase the shelf life of the food.

This Special Issue, entitled "Rapid Nondestructive Testing Technology-Based Biosensors for Food Analysis," includes ten research articles covering electrochemical sensors, colorimetric biosensors, near-infrared (NIR) spectroscopy, fluorescence spectroscopy (FS), and hyperspectral imaging (HSI). By integrating sensor data with chemometric algorithms, researchers have been able to achieve precise detection of various food quality, safety, and authenticity parameters, including freshness, nutritional components, authenticity, pesticide residue, heavy metals, fungal infection, and browning due to physiological factors.

Electrochemical sensors are devices that use the interaction between a chemical species and an electrode to produce an electrical signal, and are powerful tools for detecting and quantifying chemical species in food analysis. Electronic nose and tongue are two types of electrochemical sensors that are commonly used in food analysis to detect and identify the volatile compounds and taste components of food samples. Ren et al. [1] used miniature NIR spectroscopy and electronic tongue sensors, combined with data fusion strategies and chemometric tools, for the taste quality assessment and prediction of multiple grades of black tea, with the ant colony optimization–support vector machine (ACO–SVM) model providing the highest classification accuracy. Zhao et al. [2] proposed a novel approach to detect fungal infections in apples using a portable electronic nose and various machine learning models, including the sparrow search algorithm–backward propagation neural network (SSA–BPNN), with a high recognition accuracy of 98.40%. The study is valuable for the application of

Citation: Yun, Y.-H.; Li, J. Rapid Nondestructive Testing Technology-Based Biosensors for Food Analysis. *Biosensors* **2023**, *13*, 521. https://doi.org/10.3390/bios13050521

Received: 26 April 2023
Accepted: 27 April 2023
Published: 6 May 2023

electronic noses in the nondestructive and rapid detection of fungal infections in apples. In addition, Peng et al. [3] presented a molecularly imprinted electrochemical sensor for the rapid detection of cyromazine residues in fruits and vegetables. The sensor demonstrated good linearity and reproducibility, with a lower limit of detection of 0.5 μmol/L. This study provides a basis for the development of methods for detecting pesticide residues in edible agricultural products. Lian et al. [4] discussed the optimization of data transmission paths for pest monitoring based on the genetic algorithm (GA), particle swarm optimization (PSO), and simulated annealing (SA). The results show that the optimized path based on PSO can use the least amount of time for data transmission. The study provides a reference for improving the transmission efficiency of agricultural pest monitoring data and developing real-time and effective pest control strategies.

Colorimetric biosensors are a type of biosensor that produces a measurable color change in response to a specific analyte. When combined with a smartphone, the camera can be used to detect and quantify the color change, making it a portable and convenient method for on-site detection of various substances in food analysis. Gu et al. [5] presented a real-time detection method for Hg^{2+} in drinking water using a smartphone as a low-cost micro-spectrometer. The detection strategy uses a colorimetric sensor based on gold nanoparticles and glutathione conjugate, with a limit of detection (LOD) of 1.2 nM. The sensor was successfully applied to different types of water samples, including natural mineral water, pure water, tap water, and river water. This study also reports on the detection of gold nanoparticles with an LOD of 0.14 μM, demonstrating the versatility of the sensor.

FS is a powerful analytical technique that has been widely used in food analysis due to its high sensitivity, selectivity, and nondestructive nature. It involves the excitation of a sample with a specific wavelength of light, followed by the emission of fluorescence light at longer wavelengths. RNT technology using FS has shown great potential for food analysis, which is a promising area that can help to ensure the safety and quality of food products. Li et al. [6] used a composite thin film made up of rhodamine B encapsulated in MOF-5 as a fluorescence sensor that responds to the volatile amines produced during the quality deterioration of pork. The study uses fluorescence spectra of the composite film combined with the partial least squares (PLS) algorithm to build quantitative and qualitative models for predicting the freshness indicator and classifying pork samples, respectively. The results indicate high accuracy in predicting and classifying the freshness of pork samples. Additionally, Li et al. [7] presented a new method for detecting biothiols based on the competitive modulation of gold nanoclusters and Hg^{2+} ions, encapsulated in a zeolite imidazole framework for predesigned aggregation-induced luminescence emission. The developed fluorescence strategy has a sensitive and specific response to trace amounts of biothiols, providing a promising method for quantifying biothiols in serum, which could promote progress in disease diagnosis.

NIR measures the absorption of light in the near-infrared range (from about 780 to 2500 nm) by a sample. It is a mature nondestructive analytical technique that combines chemometrics and has become a widely used analytical tool in the food industry. Bian et al. [8] proposed a weighted multiscale support vector regression method based on variational mode decomposition (VMD-WMSVR) for the UV–Vis spectral determination of rapeseed oil adulterants and the NIR spectral quantification of Rhizoma Alpiniae Offcinarum adulterants. The VMD-WMSVR method decomposes each spectrum into discrete mode components by VMD, then builds sub-models between each component and target value using SVR, and integrates the predictions of the sub-models by weighted average to obtain the final prediction. The proposed method shows potential in model accuracy compared with PLS and SVR. Hao et al. [9] combined visible–near infrared (Vis–NIR) spectroscopy with a 1D-CNN deep learning model for online detection of browning in Yali pears. The method achieved 100% prediction accuracy for healthy and browned pears in the test set. The results indicate that the combination of Vis–NIR spectroscopy and the 1D-CNN discriminant model can be used for online detection of browning in Yali pears.

Hyperspectral imaging technology provides both spatial and spectral information about a sample, which allows for more comprehensive analysis and visualization of the sample's properties. This technology is more popular than NIR spectroscopy in many fields. Dong et al. [10] proposed a method to predict the content of alanine in beef quickly and nondestructively using near-infrared hyperspectral imaging (NIR-HSI) combined with two-dimensional correlation spectroscopy (2D-COS) analysis. The study identifies local sensitive variables related to Ala content by analyzing the sequence of chemical bond changes caused by synchronous and asynchronous correlation spectrum changes in 2D-COS. Simplified linear, nonlinear, and artificial neural network models were developed, and the PLSR model based on effective wavelengths was found to be the most effective. The results demonstrate that 2D-COS combined with NIR-HSI can be used as an effective method to monitor Ala content in beef.

Out of the ten papers published in this Special Issue, it is evident that all of them utilized chemometrics to process the analytical signal generated by various sensors, analyze large datasets, and extract valuable information. Moreover, machine learning algorithms were used to achieve precise qualitative and quantitative analysis. Notably, some researchers employed AI algorithms such as CNN to develop discriminant models. Therefore, in addition to developing high-performance sensor equipment, integrating the latest AI algorithms for deep mining of large sensor data can significantly benefit RNT technology-based biosensors for food analysis.

Acknowledgments: The authors are grateful for the opportunity to serve as the guest editors of the Special Issue, "Rapid Nondestructive Testing Technology-Based Biosensors for Food Analysis," as well as the contribution of all the authors to this Special Issue. The dedicated work of the Special Issue Editor of *Biosensors*, and the editorial and publishing staff of *Biosensors* is greatly appreciated.

Conflicts of Interest: The authors declare no conflict of interest.

References

1. Ren, G.; Zhang, X.; Wu, R.; Yin, L.; Hu, W.; Zhang, Z. Rapid Characterization of Black Tea Taste Quality Using Miniature NIR Spectroscopy and Electronic Tongue Sensors. *Biosensors* **2023**, *13*, 92. [CrossRef] [PubMed]
2. Zhao, C.; Ma, J.; Jia, W.; Wang, H.; Tian, H.; Wang, J.; Zhou, W. An Apple Fungal Infection Detection Model Based on BPNN Optimized by Sparrow Search Algorithm. *Biosensors* **2022**, *12*, 692. [CrossRef] [PubMed]
3. Peng, S.; Wang, A.; Lian, Y.; Jia, J.; Ji, X.; Yang, H.; Li, J.; Yang, S.; Liao, J.; Zhou, S. Technology for Rapid Detection of Cyromazine Residues in Fruits and Vegetables: Molecularly Imprinted Electrochemical Sensors. *Biosensors* **2022**, *12*, 414. [CrossRef] [PubMed]
4. Lian, Y.; Wang, A.; Peng, S.; Jia, J.; Zong, L.; Yang, X.; Li, J.; Zheng, R.; Yang, S.; Liao, J.; et al. Optimization of Sensors Data Transmission Paths for Pest Monitoring Based on Intelligent Algorithms. *Biosensors* **2022**, *12*, 948. [CrossRef] [PubMed]
5. Gu, Y.; Jiao, L.; Cao, F.; Liu, X.; Zhou, Y.; Yang, C.; Gao, Z.; Zhang, M.; Lin, P.; Han, Y.; et al. A Real-Time Detection Method of Hg^{2+} in Drinking Water via Portable Biosensor: Using a Smartphone as a Low-Cost Micro-Spectrometer to Read the Colorimetric Signals. *Biosensors* **2022**, *12*, 1017. [CrossRef] [PubMed]
6. Li, J.; Zhang, N.; Yang, X.; Yang, X.; Wang, Z.; Liu, H. RhB@MOF-5 Composite Film as a Fluorescence Sensor for Detection of Chilled Pork Freshness. *Biosensors* **2022**, *12*, 544. [CrossRef] [PubMed]
7. Li, S.; Wan, Y.; Li, Y.; Liu, J.; Pi, F.; Liu, L. A Competitive "On-Off-Enhanced On" AIE Fluorescence Switch for Detecting Biothiols Based on Hg^{2+} Ions and Gold Nanoclusters. *Biosensors* **2023**, *13*, 35. [CrossRef] [PubMed]
8. Bian, X.; Wu, D.; Zhang, K.; Liu, P.; Shi, H.; Tan, X.; Wang, Z. Variational Mode Decomposition Weighted Multiscale Support Vector Regression for Spectral Determination of Rapeseed Oil and Rhizoma Alpiniae Offcinarum Adulterants. *Biosensors* **2022**, *12*, 586. [CrossRef] [PubMed]
9. Hao, Y.; Li, X.; Zhang, C.; Lei, Z. Online Inspection of Browning in Yali Pears Using Visible-Near Infrared Spectroscopy and Interpretable Spectrogram-Based CNN Modeling. *Biosensors* **2023**, *13*, 203. [CrossRef] [PubMed]
10. Dong, F.; Bi, Y.; Hao, J.; Liu, S.; Lv, Y.; Cui, J.; Wang, S.; Han, Y.; Rodas-González, A. A Combination of Near-Infrared Hyperspectral Imaging with Two-Dimensional Correlation Analysis for Monitoring the Content of Alanine in Beef. *Biosensors* **2022**, *12*, 1043. [CrossRef] [PubMed]

biosensors

MDPI

Article

An Impedance-Based Immunosensor for the Detection of Ovalbumin in White Wine

Alessia Calabrese [1], Alessandro Capo [1], Angela Capaccio [1], Elettra Agovino [1], Antonio Varriale [1], Michelangelo Pascale [2], Sabato D'Auria [3,*] and Maria Staiano [2]

[1] Institute of Food Science, URT-CNR, 80126 Naples, Italy; alessia.calabrese@isa.cnr.it (A.C.); alessandro.capo@isa.cnr.it (A.C.); angela.capaccio@isa.cnr.it (A.C.); e.agovino95@libero.it (E.A.); antonio.varriale@isa.cnr.it (A.V.)
[2] Institute of Food Science, CNR, 83100 Avellino, Italy; michelangelo.pascale@cnr.it (M.P.); maria.staiano@isa.cnr.it (M.S.)
[3] Department of Biology, Agriculture and Food Science, CNR, Piazzale Aldo Moro, 7, 00185 Rome, Italy
* Correspondence: sabato.dauria@cnr.it; Tel.: +39-3683422770

Abstract: Food allergies are an exceptional response of the immune system caused by the ingestion of specific foods. The main foods responsible for allergic reactions are milk, eggs, seafood, soy, peanuts, tree nuts, wheat, and their derived products. Chicken egg ovalbumin (OVA), a common allergen molecule, is often used for the clarification process of wine. Traces of OVA remain in the wine during the fining process, and they can cause significant allergic reactions in sensitive consumers. Consequently, the European Food Safety Authority (EFSA) and the American Food and Drug Administration (FDA) have shown the risks for allergic people to assume allergenic foods and food ingredients, including eggs. Commonly, OVA detection requires sophisticated and time-consuming analytical techniques. Intending to develop a faster assay, we designed a proof-of-concept non-Faradaic impedimetric immunosensor for monitoring the presence of OVA in wine. Polyclonal antibodies anti-OVA were covalently immobilised onto an 11-mercaptoundecanoic-acid (11-MUA)-modified gold surface. The developed immunosensor was able to detect OVA in diluted white wine without the need for an external probe or any pre-treatment step with a sensitivity of 0.20 μg/mL, complying with the limit established by the resolution OIV/COMEX 502–2012 for the quantification of allergens in wine.

Keywords: allergen; wine; ovalbumin (OVA); molecular recognition element (MRE); electrochemical impedance spectroscopy (EIS); food safety; immunosensor; biosensors

Citation: Calabrese, A.; Capo, A.; Capaccio, A.; Agovino, E.; Varriale, A.; Pascale, M.; D'Auria, S.; Staiano, M. An Impedance-Based Immunosensor for the Detection of Ovalbumin in White Wine. *Biosensors* **2023**, *13*, 669. https://doi.org/10.3390/bios13070669

Received: 18 May 2023
Revised: 17 June 2023
Accepted: 20 June 2023
Published: 22 June 2023

1. Introduction

Food allergies are an abnormal body immunological reaction to the ingestion of specific foods. Milk of cows, eggs, shellfish, soy, peanuts, tree nuts, and wheat are the most-common allergenic foods. The most-efficient approach to prevent allergy is to avoid the dietary intake of such foods. However, the ubiquity of derivate food components in cooked or manufactured food products makes it challenging to avoid allergy reactions. Moreover, the body exposure to these food components may lead to a different level of allergy, from low symptoms (rush, asthma) to severe conditions such as anaphylactic responses. In this context, egg-derived products containing allergic proteins are widely employed in winemaker processing. Chicken egg ovalbumin (OVA) is one of the main proteins used in winemaking in the fining process [1]. This procedure allows for the clearing and the biochemical stabilisation of wine through the precipitation of tiny floating particles, bacteria, tartrates, proteins, pectins, various tannins, and other phenolic compounds [2]. The fining proteins can be removed by decantation, filtration steps, or a secondary fining procedure using inorganic agents, such as bentonite [3]. This operation should lead to a non-allergenic product, even if the risk for wine consumers that are allergic to egg proteins cannot be entirely excluded. An opinion paper by the European Food Safety Authority (EFSA) has highlighted the potential risks for allergic people to assume traces of OVA remained in wine during the fining process, causing significant allergic reactions [4]. Moreover, the Food and Drug Administration (FDA) in the USA has regulated the utilisation of egg products to protect consumers from the possible consequences of accidental egg consumption [5].

Within the European Union, wine producers must specify in the ingredient label the presence of allergenic aids from animal origin added for the fining process, as regulated by Directive 2007/68/EC [6]. As a result, developing accurate and sensitive approaches for detecting egg proteins in food matrices represents an opportunity for producers to identify the risk associated with residual allergens. To offset the lack of reference analytical methods for detecting fining agent proteins, the International Organization of Vine and Wine (OIV) established the analytical requirements for methods under development [7]. More precisely, the enzyme-linked immunosorbent assay (ELISA) must fulfil a detection limit of $\leq$0.25 μg/mL. This technique is the most-routinely used for monitoring proteins as allergenic compounds in food samples. However, besides requiring experienced personnel, in some cases, the ELISA test needs more sensitivity for allergen detection in some food matrices [8]. On the contrary, liquid chromatography coupled with mass spectrometry (LC-MS and LC-MS/MS) allows for complete and simultaneous identification of the analytes, representing a robust technique for detecting residual fining agents [9–12]. Nonetheless, this technique relies on time-consuming sample preparation, complicated data analysis, and costly instrumentation. In this context, biosensors might help to assess the residual presence of allergens allowing in situ and real-time food safety monitoring. A recent review reported the most-relevant progress in biosensor development for food-allergen management [13]. To date, several biosensors to detect food allergens have been proposed [14–19]. They show LoD values ranging from pg/mL to μg/mL and only a few of them have been tested on wines [18,19].

Electrochemical-impedance-based biosensors present various advantages, such as the affordability and the simplicity of miniaturisation. In particular, non-Faradaic-based biosensors, also known as capacitive sensors, detect the capacitance variations at the electrode surface produced by the molecular binding events. These sensors are well-suited to point-of-care applications thanks to their potential, such as rapid response, portability, and ease of use. Furthermore, unlike Faradaic biosensors, which require adding a potentially harmful redox probe, which could damage biomolecules [20], the non-Faradaic approach is label-free, allowing the analysis without sample preparation or trained personnel [21,22].

In this work, we explored the application of electrochemical impedance spectroscopy (EIS) to develop a non-Faradaic impedimetric immunosensor for detecting the presence of OVA in white wine. For this purpose, polyclonal antibodies against OVA (pAb anti-OVA) were produced and characterised by an indirect ELISA test. Then, the antibodies were covalently attached to a gold-based electrode surface. Atomic force microscopy (AFM) was employed to study the surface topography of the functionalised biosensor surfaces. Finally, EIS was used to characterise the performance of the immunosensor, which allowed for a simple and fast determination of OVA directly in diluted white wine without any additional reagent.

2. Materials and Methods

2.1. Reagents and Instruments

All commercially available reagents were chosen for the highest quality. 11-mercaptoundecanoic acid (11-MUA), N-hydroxysuccinimide (NHS), N-(3-dimethylaminopropyl)-N′-ethylcarbodiimide hydrochloride (EDC), sulfuric acid (H_2SO_4), ethanolamine (ETA), ethanol, albumin from chicken egg (OVA), and 3,5-tetramethylbenzidine (TMB) were purchased from Sigma-Aldrich (Milan, Italy). Goat polyclonal to rabbit IgG-HRP conjugate (secondary antibody) was from Abcam (Cambridge, U.K.). The antibodies against OVA were produced and purchased from Covalab SAS (Bron, France). nProtein A Sepharose 4 Fast Flow resin for antibody purification was acquired from Cytiva (Washington, DC,

USA). Materials used for protein electrophoresis were obtained from Bio-Rad (Hercules, CA, USA). Microplates (96-well), Nunc LockWell MaxiSorp form, and a microplate reader, Tecan Infinity 200 Pro (Tecan, Männedorf, Switzerland), were used for the indirect ELISA. UV measurements were carried out on a Jasco V-730 UV/Vis spectrophotometer. Ultrafiltration units Vivaspin® Turbo 15 (30,000 MWCO) for antibody concentration were from Sartorius (Göttingen, Germany). The miniaturised All-in-One electrochemical workstation (MicruX ECStat), the All-in-One cell, the batch-cell Add-on, and thin-film gold single electrodes (ED-SE1-Au) were purchased from Micrux Technologies (Oviedo, Spain). White wine (alcohol content: 10.5%) was acquired from a local market.

2.2. *Antibody Production and Purification*

Antibodies anti-OVA were produced by Covalab (France), according to Varriale et al. (2016) [23]. In brief, two rabbits were immunised following a standard protocol by intradermal inoculation of an antigen (0.5 mg of OVA per rabbit). After the immunisation period, the rabbits were sacrificed. Their blood was recovered and centrifuged to separate the blood cells from the serum. From the obtained serum, the antibodies were purified, according to Pennacchio et al. (2016) [24]. In particular, 1 mL of rabbit serum diluted with 1.0 mL of sodium phosphate (NaP) 20 mM, pH 7.0, was applied to a Protein A column. Then, the IgG fraction was eluted with sodium citrate 0.1 M, pH 3.0, and immediately buffered in sodium borate 1 M, pH 8.5. Finally, the IgG concentration and purity were checked by absorbance measurement at 278 nm and SDS-PAGE (12% acrylamide), respectively. The obtained pure samples were concentrated at 1.6 mg/mL by centrifugal concentrators.

2.3. *Indirect ELISA*

In order to verify the binding capacity of the produced antibodies, an indirect ELISA was performed, according to Capo et al. (2022) [25]. The antigen OVA was dissolved in carbonate buffer (coating buffer) 0.05 M, pH 9.6, and diluted from 0.005 μg/mL to 50 μg/mL. The 50 μL/well of each dilution was used to coat 96-well microplates, incubated overnight at 4 °C. Coating buffer and bovine serum albumin (BSA) (100 μg/mL) were used as controls. The plate was rinsed thrice with TBS-T (TBS 0.01 M, pH 7.4; 0.05% Tween-20), incubated with 200 μL/well of blocking buffer (TBS; 5% *w*/*v* non-fat dried milk) at 37 °C for 2 h, and rewashed three times with TBS-T. Afterward, pAb anti-OVA 1 μg/mL (50 μL/well) diluted in blocking buffer (TBS; 1% non-fat dried milk; 0.05% *v*/*v* Tween-20; pH 7.4) was incubated at 37 °C for 2 h. After three steps of washing with TBS-T, 50 μL/well of horseradish-peroxidase (HRP)-conjugated goat anti-rabbit IgG antibodies (0.5 μg/mL) was diluted in blocking buffer (TBS; 1% non-fat dried milk; 0.05% *v*/*v* Tween-20; pH 7.4) and incubated for 1 h at 37 °C. After an incubation of 10 min at 37 °C with the TMB substrate (100 μL/well), the stopping solution (HCl 2.5 M; 50 μL/well) was added to stop the colour development, and the absorbance was recorded at 450 nm.

2.4. *Immunosensor Development*

The electrochemical sensors used in this work were gold-based and consisted of a three-electrode configuration (reference, working, and counter, as shown in Figure 1 from the left to the right of the chip). Before the derivatisation procedure, an electrochemical surface pre-cleaning was performed through 12 potential cycles in the range of −1.0 to +1.3 V, at a scan rate of 0.1 V/s, in the presence of sulfuric acid (H_2SO_4) (0.05 M (5 μL)). Next, the clean gold substrates were immersed in an ethanolic solution of 11-MUA 5 mM (350 μL) for 24 h [26]. Afterward, the antibodies were immobilised on the working electrode via carbodiimide-mediated coupling in two consecutive steps: (1) a 10 min incubation with a mixture of EDC/NHS (5 μL) (50 mM/5 mM) (volume ratio 1:1) in phosphate-buffered saline (PBS, pH 7.4); (2) a 2 h incubation with pAb anti-OVA solution (0.25 mg/mL (5 μL)). Lastly, the unreacted active sites were blocked with ethanolamine 1 M (pH 8.5, 5 μL) for 20 min (Figure 1).

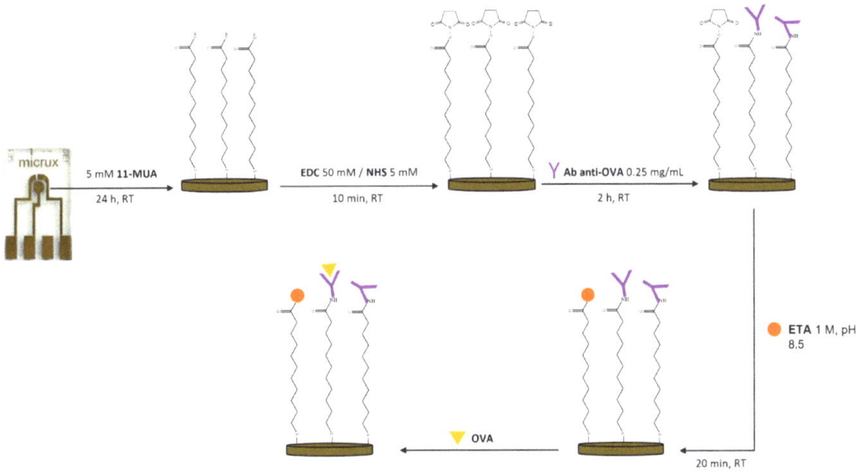

Figure 1. Schematic diagram of the immunosensor fabrication. Gold thin-film electrochemical sensors (ED-SE1-Au) were treated sequentially with 11-MUA, a mixture of NHS and EDC, a solution of pAb anti-OVA, and ETA. The immunosensor was tested at different concentrations of OVA.

2.5. Surface Characterisation by Atomic Force Microscopy

The immobilisation of the pAb anti-OVA on the gold impedimetric surface was characterised by AFM. The AFM images were acquired using a Ntegra Prima system (NT-MDT Spectrum Instruments, Zelenograd Russia) equipped with an AFM silicon tip (NSG01, NT-MDT) having a spring constant of 5 N/m and a nominal resonance frequency of 160 KHz in air. AFM measurements were carried out in tapping mode and in the PBS buffer to preserve the structure and function of the biological components. For each sample, random 1 μm × 1 μm maps were scanned throughout the working area of the gold electrode with a resolution of 200 pixels per line and a scan rate of 2 Hz. From the topographic images, the root-mean-squared roughness (RMS) was calculated and averaged on ten AFM scans after image processing with the open-source software Gwyddion 2.62.

2.6. Impedimetric Measurements

Non-Faradaic impedance spectroscopy is able to investigate antigen–antibody biorecognition events without the use of an external redox probe.

The measuring cell used consisted of an All-in-One platform, which enabled the use of the thin-film electrodes in static conditions, with a batch-cell Add-on, which provided an interface with the electrochemical workstation, and it facilitated the dropping of the sample on the electrode.

For this purpose, the working area of the immunosensor was incubated for 10 min with 5 μL of OVA dissolved in PBS (pH 7.4) at different concentrations (0.001, 0.005, 0.01, 0.05, 0.1, and 0.5 μg/mL).

After thoroughly washing the surfaces with PBS, impedimetric measurements were performed in PBS at 25 °C, using a sinusoidal AC potential (0.1 V) and a DC potential of 0 V in the frequency range of 0.1 to 100,000 Hz. The AC potential value was chosen after analysing the impedance response of the system at different values (0.005, 0.01, 0.05, and 0.1 V) (Figure S1). To test the cross-reactivity of the immunosensor, BSA at increasing concentrations was used as a negative control (Figure S2).

2.7. Preparation of Wine Samples for Impedimetric Tests on Real Matrix

In order to test the performance of the immunosensor on a real matrix, a white wine acquired in a local grocery was diluted 1:200 in PBS (pH 7.4) and was spiked with different concentrations of OVA (0.001, 0.005, 0.01, 0.05, 0.1, and 0.5 μg/mL). The dilution ratio was chosen to minimise matrix effects, preserving the characteristics of the immunosensor [18]. The working area of the electrode was incubated for 10 min with 5 μL of the sample. After extensive washing steps, impedimetric measurements were performed as previously described.

2.8. Statistical Analysis

All the measurements were carried out in triplicate.

The standard deviation (SD) for all the data reported was calculated from the formula:

$$SD = \sqrt{\frac{\sum(x-\bar{x})^2}{(n-1)}}, \quad (1)$$

where x is the sample mean average and n is the sample size.

The limit of detection (LoD) for indirect ELISA was calculated according to Ambruster et al. (2008) (LoD = LoB + 1.645 (SD low concentration OVA) [27] and for the immunosensor performance following Shrivastava (2011) (LoD = 3.3 S/b) [28]. Data were analysed in Microsoft® Excel 2016 and Origin® 2018.

3. Results and Discussion

In this work, pAb anti-OVA were used to develop a non-Faradaic impedimetric immunosensor to detect the presence of traces of ovalbumin in white wine.

3.1. Evaluation of the Antibody-Binding Capacity by Indirect ELISA

After the Ab purification procedure (described in Section 2.2), the binding capacity of the pAb anti-OVA was assessed by an indirect ELISA test. Figure 2 shows that the pAbs anti-OVA were able to recognise the antigen up to 0.05 μg/mL. The limit of detection (LoD), determined by considering the measured limit of blank (LoB) and the replicates (n = 3) of the sample containing a low concentration of analyte LoD = LoB + 1.645 (SD low-concentration OVA) [27], was estimated to be 0.09 μg/mL.

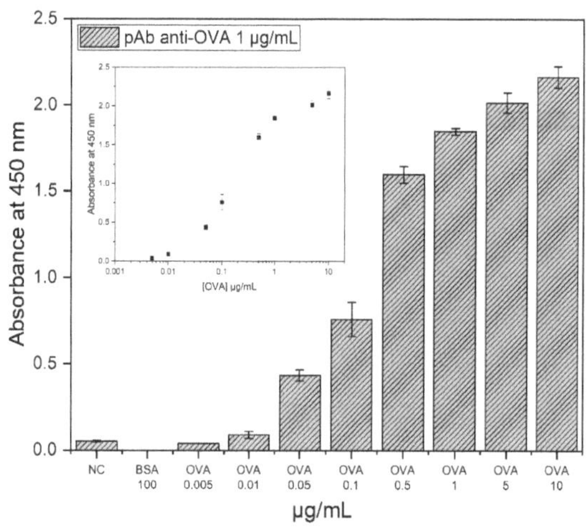

Figure 2. Indirect ELISA results for pAb anti-OVA (1 μg/mL) binding capacity. The figure shows that OVA was recognised by the pAbs anti-OVA up to 0.05 μg/mL (NC: no coating; BSA used as a negative control). The insert shows the variation of signal at 450 nm as function of the ovalbumin concentration.

3.2. Electrochemical Characterisation of the Immunosensor Assembling

A typical biorecognition surface for capacitance detection comprises two layers: a double insulation layer and a recognition layer; a third layer is generated by analyte binding to the recognition element. A high-capacitance biorecognition layer detects slight variations induced by the binding event [29,30]. Thus, achieving a tightly packed biorecognition layer (SAM: self-assembled monolayer of long thiols C11–C16) allows for maximising the ability to detect small impedance and capacity variations, reducing unrestricted ionic migration at the interface.

The pAbs anti-OVA were covalently immobilised on an 11-MUA-modified gold surface SAM (Section 2.4). The electrode derivatisation for the immunosensor assembling was monitored using non-Faradaic electrochemical impedance spectroscopy (EIS), characterised by resistive and capacitive contributions. A non-Faradaic system directly detects the analyte without needing a redox probe, such as a ferrocyanide–ferricyanide redox couple ($Fe\,(CN)_6{}^{3-/4-}$). As a result, the experimental process turns out to be more manageable and suitable for rapid monitoring.

The impedance data are presented in the Nyquist plot, where the imaginary part (Z'') is plotted versus the real part (Z') of an impedance Z over a specified frequency range (0.1 to 100,000 Hz). Due to the absence of a redox probe, a non-Faradaic Nyquist plot shows a large incomplete semicircle, lacking the parameters related to electron transfer, including charge transfer resistance (R_{ct}) and Warburg impedance (Z_W). Therefore, the impedance of a non-Faradaic sensor is determined by the insulating characteristics of the species bond to the conductive substrate. As a result, the deposition of the consecutive layers on the electrode causes an increase in the impedance of the system (Figure 3). This variation could be attributed to the coating layer on the electrode surface, which increases during the assembly steps.

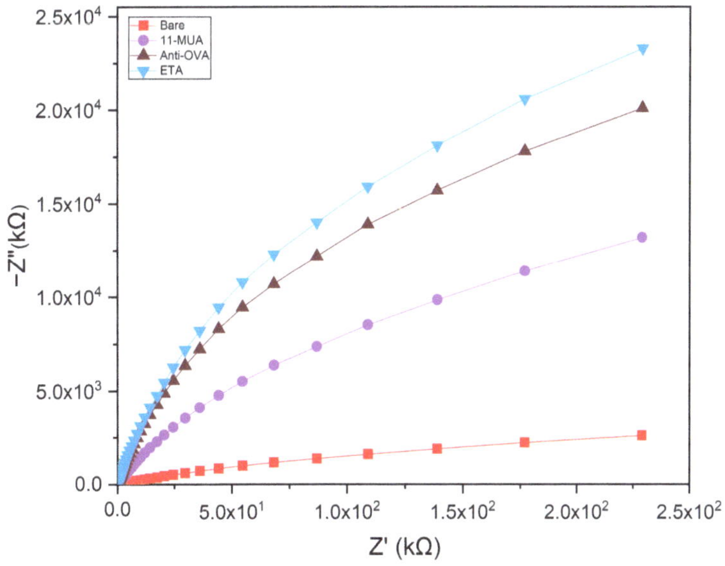

Figure 3. Nyquist plots for the different functionalisation steps performed for the fabrication of the impedimetric immunosensor. The impedance of the electrochemical system increased after each step of the functionalisation procedure.

3.3. AFM Analysis

Since the electrochemical response depends on the morphological characteristics of the surface electrode, AFM is useful to evaluate the immobilisation of the pAbs anti-OVA on the impedimetric immunosensor. In this respect, the root mean square (RMS) value of the height irregularities was chosen to characterise the immunoassay surface roughness [31]. The following AFM analysis was carried out on the same chip at different stages of the pAb anti-OVA immobilisation process to avoid the fluctuations of the surface roughness due to the inter-batch variability of the electrochemical surface sensors. Figure 4a illustrates the AFM morphology map of 1 μm × 1 μm in size acquired on the working area of the bare gold-based electrode. The surface of the electrochemical sensor was characterised by a low roughness (RMS = 1.7 ± 0.1 nm) and showed the typical granular structure, consisting of nearly spherical nanoparticles, which was strictly connected to the gold deposition process parameters used by Micrux manufactory [32]. The sensor was, then, functionalised with an 11-MUA SAM, as described in Section 2.4. The evidence that the gold surface was chemically modified was emphasised by the significant change of the surface roughness (RMS = 2.3 ± 0.1 nm) as observed in the AFM morphology of the surface (Figure 4b). Finally, in Figure 4c is shown the AFM morphology map of the 11-MUA-modified gold surface after the incubation of the pAbs anti-OVA. As highlighted in the bar plot of Figure 4d, the deposition of the consecutive layers on the electrode affected the RMS value of the surfaces. In particular, the increment of the RMS value of the pAb anti-OVA surface (2.6 ± 0.1 nm) compared to the 11-MUA-modified gold surface and the bare electrode surface was comparable to the roughness change detected in similar studies [33,34]. The AFM results further assessed the complete immobilisation process of the pAbs anti-OVA as also confirmed by the impedimetric experiments.

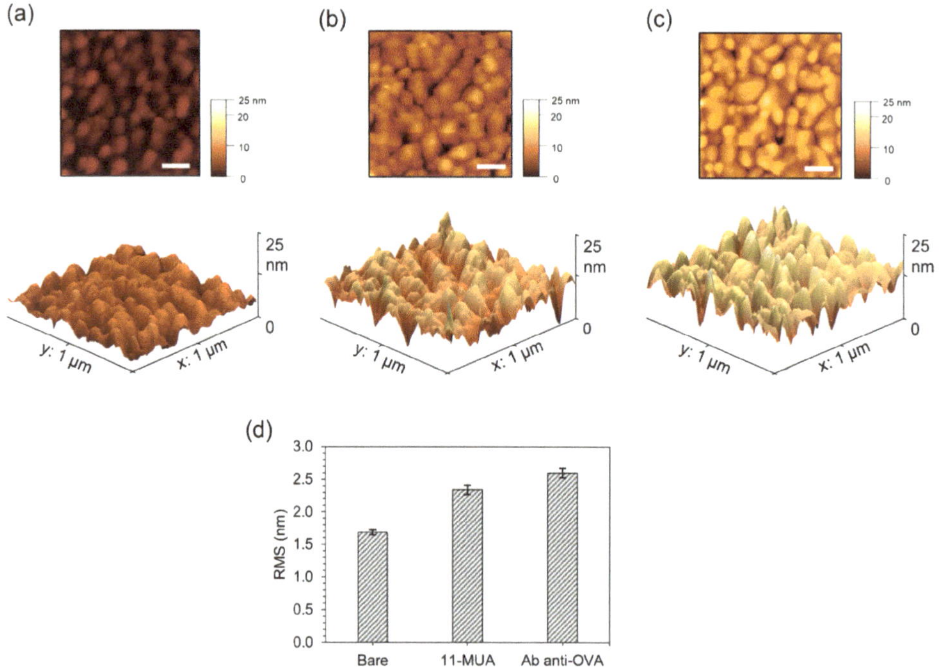

Figure 4. AFM analysis of pAb anti-OVA immobilisation on the electrochemical immunosensor. The 2D and 3D AFM morphology scans (1 μm × 1 μm) acquired in tapping mode of (**a**) the bare gold electrode, 11-MUA-modified gold surface (**b**) before and (**c**) after pAb anti-OVA incubation. The white scale bar in the 2D image corresponds to 20 nm. (**d**) Bar plot of the root-mean-squared roughness (RMS) of each analysed surface and presented as the average ± the SD (n = 10).

3.4. Electrochemical Characterisation of the Immunosensor Performance

The electrochemical performance of the immunosensor was examined by non-Faradaic EIS for detecting OVA dissolved in PBS. Unlike the Faradaic approach, a non-Faradaic response shows high impedance values given that no redox species assist the charge transfer between the interfacial layers [35]. The impedance response was recorded starting from the lowest (0.001 μg/mL) to the highest (0.5 μg/mL) concentration of OVA. The insulating effect of the sensing surface and the OVA binding on the modified electrode generated variations in the double-layer capacitance (C_{dl}). The binding phenomena occurring at the electrode interface can be observed in the Nyquist plot as an increase in the imaginary part of impedance at increasing OVA concentrations (Figure 5).

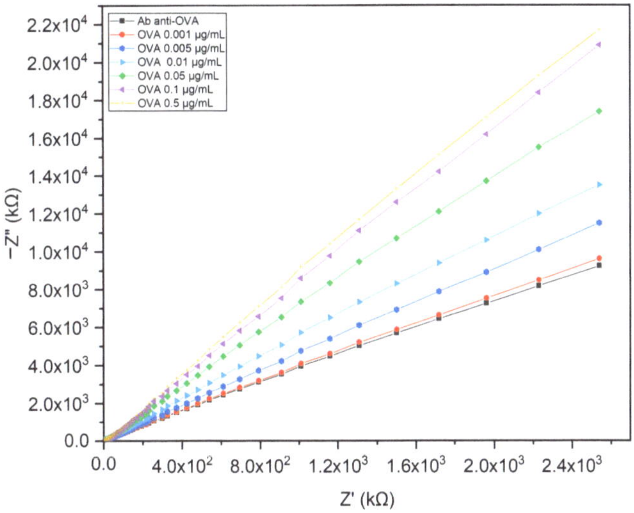

Figure 5. Nyquist plot of the pAb-anti-OVA-based immunosensor tested on OVA in PBS. The impedance of the electrochemical system increased at increasing concentrations of OVA.

As the maximum impedance variation was observed at 0.1 Hz, the impedance values registered at this frequency allowed us to obtain the binding curve described by a non-linear fitting function. Figure 6a shows the plot of the change in impedance at 0.1 Hz expressed as ΔZ ($Z_{OVA} - Z_{blank}$) versus OVA concentration.

A linear correlation was observed in the range of 0.001 to 0.01 μg/mL, while for higher concentrations (from 0.05 to 0.5 μg/mL), no linear trend was noticeable, due to the saturation of the binding sites. Therefore, the calibration curve was calculated in the linear range to determine the detection limit (LoD) of the immunosensor in PBS (Figure 6b). Each point represents the average of three replicates, and the error bars are the standard deviations of the mean (SDs $\leq$ 380 ΔZ). The LoD, calculated by 3.3 S/b, where S is the standard deviation of the y-intercept of the linear regression and b is the slope of the linear range [27], was estimated to be 0.0008 μg/mL, with a response time of 15 min, including the incubation time. As a control, the surface was tested at increasing concentrations of BSA, and no significant impedance variations were registered (Figure S2).

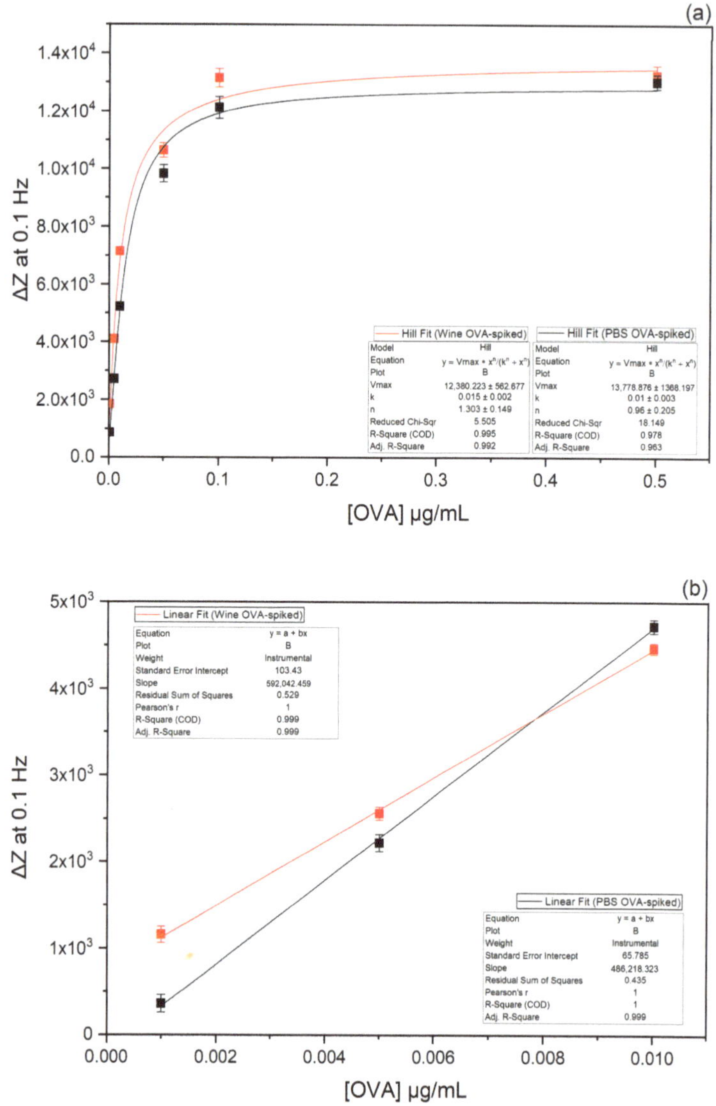

Figure 6. Plots of the change in impedance (ΔZ) at 0.1 Hz versus OVA concentration dissolved in PBS (in black colour) and OVA concentration in spiked white wine (in red colour). The binding curves were obtained through a non-linear fitting function (**a**). The calibration curves of the immunosensor were obtained through a linear fitting function (**b**).

3.5. Electrochemical Characterisation of the Immunosensor Performance on Real Matrix

In order to explore the application of the immunosensor on a real matrix, impedance measurements were conducted on white wine diluted in PBS and spiked with different concentrations of OVA (0.001, 0.005, 0.01, 0.05, 0.1, and 0.5 μg/mL) as described in Section 2.7. The impedimetric response of the electrode reported in Figure 7 displays that the immunosensor was able to detect OVA even in wine.

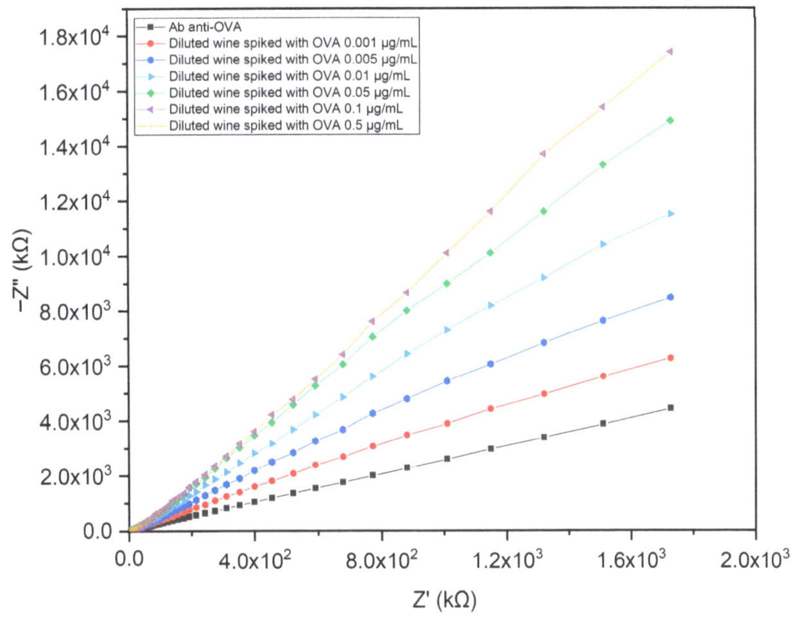

Figure 7. Nyquist plot of the pAb-anti-OVA-based immunosensor tested with white wine diluted in PBS and spiked with OVA.

Taking into consideration that the maximum impedance variation was at 0.1 Hz, the impedance values registered at this frequency were used to obtain the binding curve by applying a non-linear fitting function. The plot of the change in impedance at 0.1 Hz expressed as ΔZ ($Z_{\text{OVA in wine}} - Z_{\text{blank}}$) versus the OVA concentration is shown in Figure 6a.

The calibration curve (Figure 6b) was calculated in the linear range of 0.001 to 0.01 μg/mL to determine the detection limit of the immunosensor in wine. The graph shows standard deviation error bars from the mean of triplicate measurements (SDs $\leq$ 340 ΔZ). The LoD, calculated by 3.3 S/b [28], was estimated to be 0.001 μg/mL, with a response time of 15 min, including the incubation time. By considering the 200-fold dilution used in the measurements, the LoD of the assay in white wine was 0.20 μg/mL, a value that is compliant with the detection limit recommended by the OIV (i.e., 0.25 μg/mL) for the quantification of allergens in wine [7].

Reproducibility data calculated in terms of the relative standard deviation (RSD) are reported in Table S1. The immunosensor, stored in PBS buffer at 4 °C, showed stability for 2 weeks after the functionalisation procedure.

Table 1 depicts a comparison of the analytical performance of the immunosensor with other biosensors for OVA detection previously reported. The detection limit obtained in this work is compatible with the limit established by the resolution OIV/COMEX 502–2012 for quantifying allergens in wine. In addition, the immunosensor presents advantages such as a rapid response, the ease of the functionalisation process, the need for micro volumes of the sample, an assay time of 15 min, and affordability.

Table 1. Performance comparison of different biosensors for OVA detection.

Sensing Substrate	Technique	Linear Range	LoD	Assay Time (min)	Number of Steps	References
GO/screen-printed carbon	Differential pulse voltammetry	1 pg/mL–0.5 μg/mL	0.83 pg/mL	60	5	Eissa et al. (2013) [15]
Dextran-coated sensor chips (CM5)	Surface plasmon resonance (SPR)	0.03–0.2 μg/mL	0.6 μg/mL	-	4	Pilolli et al. (2015) [16]
Screen-printed platinum	Linear sweep voltammetry	0.5–9.5 μg/mL	0.2 μg/mL	60	4	Čadková et al. (2015) [17]
Graphene/screen-printed carbon electrodes	Amperometry	0.01–10 pg/mL	0.2 fg/mL	60	6	Baldo et al. (2021) [18]
Thin-film gold single electrodes	Non-Faradaic impedance spectroscopy	0.001–0.01 μg/mL	0.2 μg/mL	15	4	This work

4. Conclusions

In this work, we presented a non-Faradaic impedimetric immunosensor to monitor the presence of OVA in white wine. The primary purpose of this study was to explore the feasibility of using an impedimetric pAb-anti-OVA-based biosensor on a real matrix without the use of external probes and the need for sample pre-treatment. The immunosensor showed a sensitivity of 0.20 μg/mL, which is compatible with the limit established by the resolution OIV/COMEX 502–2012 for quantifying the presence of allergens in wine. Although the developed biosensor presents some advantages with respect to the biosensors for OVA detection present in the literature (such as affordability, a good response time, and the direct monitoring of OVA in diluted samples), additional efforts will be devoted to improving the performance of the biosensor.

Supplementary Materials: The following Supporting Information can be downloaded at: https://www.mdpi.com/article/10.3390/bios13070669/s1, Figure S1: EIS spectra acquired at different AC potentials in the frequency range of 0.1 to 100,000 Hz.; Figure S2: Immunosensor BSA cross-reactivity.; Table S1: Immunosensor reproducibility data.

Author Contributions: Conceptualisation, M.S., A.V., M.P. and S.D.; methodology, A.C. (Alessia Calabrese), A.C. (Alessandro Capo), E.A. and A.C. (Angela Capaccio); software, A.C. (Alessia Calabrese) and A.C. (Alessandro Capo); validation, A.C. (Alessandro Capo) and A.V.; formal analysis, A.C. (Alessia Calabrese) and A.C. (Alessandro Capo); investigation, A.C. (Alessia Calabrese), E.A., A.C. (Alessandro Capo) and A.C. (Angela Capaccio); resources, M.S.; data curation, A.C. (Alessia Calabrese); writing—original draft preparation, A.V., A.C. (Alessia Calabrese), M.S. and S.D.; writing—review and editing, A.C. (Alessandro Capo), A.V., M.S., M.P. and S.D.; visualisation, A.C. (Alessia Calabrese); supervision, M.S., A.V. and S.D.; project administration, S.D. and M.S.; funding acquisition, S.D. All authors have read and agreed to the published version of the manuscript.

Funding: This study was carried out within the AGRITECH Spoke 9—Codice progetto MUR: AGRITECH "National Research Centre for Agricultural Technologies"—CUP CN00000022, a valere sulla Missione 4 Componente 2 (M4C2)—Investimento 1.4 "Potenziamento strutture di ricerca e creazione di "campioni nazionali di R&S" su alcune Key Enabling Technologies" del Piano Nazionale di Ripresa e Resilienza (PNRR) finanziato dall'Unione Europea "Next Generation EU".

Institutional Review Board Statement: Not applicable.

Informed Consent Statement: Not applicable.

Data Availability Statement: The data presented in this study are available within the article and its Supplementary Materials. Other data that support the findings of this study are available upon request from the corresponding author.

Conflicts of Interest: The authors declare no conflict of interest.

References

1. Mine, Y.; Yang, M. Recent advances in the understanding of egg allergens: Basic, industrial, and clinical perspectives. *J. Agric. Food Chem.* **2008**, *56*, 4874–4900. [CrossRef]
2. Ghanem, C.; Taillandier, P.; Rizk, M.; Rizk, Z.; Nehme, N.; Souchard, J.P.; El Rayess, Y. Analysis of the impact of fining agents types, oenological tannins and mannoproteins and their concentrations on the phenolic composition of red wine. *LWT Food Sci. Technol.* **2017**, *83*, 101–109. [CrossRef]

3. Weiss, K.; Bisson, L. Effect of Bentonite Treatment of Grape Juice on Yeast Fermentation. *Am. J. Enol. Vitic.* **2002**, *53*, 28–36. [CrossRef]
4. EFSA Panel on Dietetic Products, Nutrition and Allergies. Scientific Opinion on the evaluation of allergenic foods and food ingredients for labelling purposes. *EFSA J.* **2014**, *12*, 3894. [CrossRef]
5. Food and Drug Administration, U.S. Department of Health and Human Services. Egg Regulatory Program Standards (ERPS). Available online: https://www.fda.gov/federal-state-local-tribal-and-territorial-officials/regulatory-program-standards/egg-regulatory-program-standards-erps (accessed on 15 May 2023).
6. European Commission. Commission Implementing Regulation (EU) no 579/2012. L171. *Off. J. Eur. Union* **2012**, *L171*, 4–7.
7. OIV. Criteria for the Quantification of Potentially Allergenic Residues of Fining Agent Proteins in Wine. OIV-OENO 427–2010 Modified by OIV–Comex 502–2012. 2012. Available online: https://www.oiv.int/public/medias/2154/oiv-comex-502-2012-en.pdf (accessed on 5 April 2023).
8. Van Hengel, A.J. Food allergen detection methods and the challenge to protect food-allergic consumers. *Anal. Bioanal. Chem.* **2007**, *389*, 111–118. [CrossRef]
9. Monaci, L.; Visconti, A. Mass spectrometry-based proteomics methods for analysis of food allergens. *TrAC Trends Anal. Chem.* **2009**, *28*, 581–591. [CrossRef]
10. Johnson, P.E.; Baumgartner, S.; Aldick, T.; Bessant, C.; Giosafatto, V.; Heick, J.; Mamone, G.; O'Connor, G.; Poms, R.; Popping, B.; et al. Current perspectives and recommendations for the development of mass spectrometry methods for the determination of allergens in foods. *J. AOAC Int.* **2011**, *94*, 1026–1033. [CrossRef] [PubMed]
11. Cryar, A.; Pritchard, C.; Burkitt, W.; Walker, M.; O'Connor, G.; Burns, D.T.; Quaglia, M. Towards absolute quantification of allergenic proteins in food--lysozyme in wine as a model system for metrologically traceable mass spectrometric methods and certified reference materials. *J. AOAC Int.* **2013**, *96*, 1350–1361. [CrossRef]
12. Monaci, L.; Pilolli, R.; De Angelis, E.; Godula, M.; Visconti, A. Multi-allergen detection in food by micro high-performance liquid chromatography coupled to a dual cell linear ion trap mass spectrometry. *J. Chromatogr. A* **2014**, *1358*, 136–144. [CrossRef]
13. Shin, J.H.; Reddy, Y.V.M.; Park, T.J.; Park, J.P. Recent advances in analytical strategies and microsystems for food allergen detection. *Food Chem.* **2022**, *371*, 131120. [CrossRef] [PubMed]
14. Ocaña, C.; Hayat, A.; Mishra, R.; Vasilescu, A.; Del Valle, M.; Marty, J.-L. A novel electrochemical aptamer–antibody sandwich assay for lysozyme detection. *Analyst* **2015**, *140*, 4148–4153. [CrossRef]
15. Eissa, S.; L'Hocine, L.; Siaj, M.; Zourob, M. A graphene-based label-free voltammetric immunosensor for sensitive detection of the egg allergen ovalbumin. *Analyst* **2013**, *138*, 4378–4384. [CrossRef] [PubMed]
16. Pilolli, R.; Visconti, A.; Monaci, L. Rapid and label-free detection of egg allergen traces in wines by surface plasmon resonance biosensor. *Anal. Bioanal. Chem.* **2015**, *407*, 3787–3797. [CrossRef]
17. Čadková, M.; Metelka, R.; Holubová, L.; Horák, D.; Dvořáková, V.; Bílková, Z.; Korecká, L. Magnetic beads-based electrochemical immunosensor for monitoring allergenic food proteins. *Anal. Biochem.* **2015**, *484*, 4–8. [CrossRef]
18. Baldo, T.A.; Proença, C.D.A.; Felix, F.D.S.; Freitas, T.A.; Sakata, S.K.; Angnes, L.; Faria, R.C. Disposable electrochemical microfluidic device for ultrasensitive detection of egg allergen in wine samples. *Talanta* **2021**, *232*, 122447. [CrossRef] [PubMed]
19. Titoiu, A.M.; Porumb, R.; Fanjul-Bolado, P.; Epure, P.; Zamfir, M.; Vasilescu, A. Detection of Allergenic Lysozyme during Winemaking with an Electrochemical Aptasensor. *Electroanalysis* **2019**, *31*, 2262–2273. [CrossRef]
20. Daniels, J.S.; Pourmand, N. Label-Free Impedance Biosensors: Opportunities and Challenges. *Electroanalysis* **2007**, *19*, 1239–1257. [CrossRef]
21. Qureshi, A.; Niazi, J.H.; Kallempudi, S.; Gurbuz, Y. Label-free capacitive biosensor for sensitive detection of multiple biomarkers using gold interdigitated capacitor arrays. *Biosens. Bioelectron.* **2010**, *25*, 2318–2323. [CrossRef]
22. Magar, H.S.; Hassan, R.Y.A.; Mulchandani, A. Electrochemical Impedance Spectroscopy (EIS): Principles, Construction, and Biosensing Applications. *Sensors* **2021**, *21*, 6578. [CrossRef]
23. Varriale, A.; Bonnot, K.; Peransi, S.; Scala, A.; D'Auria, S. Self-oriented monolayer immobilization of ovalbumin and B. cereus antibody molecules on a chemically modified surface of silicon nitride fosters the enhancement of capture of bio-agents. *Colloids Surf. B Biointerfaces* **2016**, *148*, 585–591. [CrossRef] [PubMed]
24. Pennacchio, A.; Varriale, A.; Scala, A.; Marzullo, V.M.; Staiano, M.; D'Auria, S. A novel fluorescence polarization assay for determination of penicillin G in milk. *Food Chem.* **2016**, *190*, 381–385. [CrossRef]
25. Capo, A.; Calabrese, A.; Frant, M.; Walczak, M.; Szczotka-Bochniarz, A.; Manessis, G.; Bossis, I.; Staiano, M.; D'Auria, S.; Varriale, A. SPR-Based Detection of ASF Virus in Cells. *Int. J. Mol. Sci.* **2022**, *23*, 7463. [CrossRef] [PubMed]
26. Ahmad, A.; Moore, E. Electrochemical immunosensor modified with self-assembled monolayer of 11-mercaptoundecanoic acid on gold electrodes for detection of benzo[a]pyrene in water. *Analyst* **2012**, *137*, 5839–5844. [CrossRef] [PubMed]
27. Armbruster, D.A.; Pry, T. Limit of blank, limit of detection and limit of quantitation. *Clin. Biochemist. Rev.* **2008**, *29* (Suppl. S1), S49–S52.
28. Shrivastava, A. Methods for the determination of limit of detection and limit of quantitation of the analytical methods. *Chron. Young Sci.* **2011**, *2*, 21–25. [CrossRef]
29. Berggren, C.; Bjarnason, B.; Johansson, G. Capacitive Biosensors. *Electroanalysis* **2001**, *13*, 173–180. [CrossRef]
30. Chen, D.; Li, J. Interfacial design and functionization on metal electrodes through self-assembled monolayers. *Surf. Sci. Rep.* **2006**, *61*, 445–463. [CrossRef]

31. Tuttle, P.V.t.; Rundell, A.E.; Webster, T.J. Influence of biologically inspired nanometer surface roughness on antigen-antibody interactions for immunoassay-biosensor applications. *Int. J. Nanomed.* **2006**, *1*, 497–505. [CrossRef]
32. Ruffino, F.; Grimaldi, M.G. Morphological Characteristics of Au Films Deposited on Ti: A Combined SEM-AFM Study. *Coatings* **2018**, *8*, 121. [CrossRef]
33. Wang, S.; Esfahani, M.; Gurkan, U.A.; Inci, F.; Kuritzkes, D.R.; Demirci, U. Efficient on-chip isolation of HIV subtypes. *Lab Chip* **2012**, *12*, 1508–1515. [CrossRef] [PubMed]
34. Zorea, J.; Shukla, R.P.; Elkabets, M.; Ben-Yoav, H. Probing antibody surface density and analyte antigen incubation time as dominant parameters influencing the antibody-antigen recognition events of a non-faradaic and diffusion-restricted electrochemical immunosensor. *Anal. Bioanal. Chem.* **2020**, *412*, 1709–1717. [CrossRef] [PubMed]
35. Katz, E.; Willner, I. Probing Biomolecular Interactions at Conductive and Semiconductive Surfaces by Impedance Spectroscopy: Routes to Impedimetric Immunosensors, DNA-Sensors, and Enzyme Biosensors. *Electroanalysis* **2003**, *15*, 913–947. [CrossRef]

biosensors

MDPI

Article

Rapid Characterization of Black Tea Taste Quality Using Miniature NIR Spectroscopy and Electronic Tongue Sensors

Guangxin Ren [1,2,3,*], Xusheng Zhang [1,4], Rui Wu [1,3], Lingling Yin [1,3], Wenyan Hu [1,3] and Zhengzhu Zhang [2]

1 School of Biological Engineering, Institute of Digital Ecology and Health, Huainan Normal University, Huainan 232038, China
2 State Key Laboratory of Tea Plant Biology and Utilization, Anhui Agricultural University, Hefei 230036, China
3 Key Laboratory of Bioresource and Environmental Biotechnology of Anhui Higher Education Institutes, Huainan Normal University, Huainan 232038, China
4 Library, Huainan Normal University, Huainan 232038, China
* Correspondence: rgx@hnnu.edu.cn

Abstract: The taste of tea is one of the key indicators in the evaluation of its quality and is a key factor in its grading and market pricing. To objectively and digitally evaluate the taste quality of tea leaves, miniature near-infrared (NIR) spectroscopy and electronic tongue (ET) sensors are considered effective sensor signals for the characterization of the taste quality of tea leaves. This study used micro-NIR spectroscopy and ET sensors in combination with data fusion strategies and chemometric tools for the taste quality assessment and prediction of multiple grades of black tea. Using NIR features and ET sensor signals as fused information, the data optimization based on grey wolf optimization, ant colony optimization (ACO), particle swarm optimization, and non-dominated sorting genetic algorithm II were employed as modeling features, combined with support vector machine (SVM), extreme learning machine and *K*-nearest neighbor algorithm to build the classification models. The results obtained showed that the ACO−SVM model had the highest classification accuracy with a discriminant rate of 93.56%. The overall results reveal that it is feasible to qualitatively distinguish black tea grades and categories by NIR spectroscopy and ET techniques.

Keywords: tea; near-infrared spectroscopy; electronic tongue; chemometrics; quality control

Citation: Ren, G.; Zhang, X.; Wu, R.; Yin, L.; Hu, W.; Zhang, Z. Rapid Characterization of Black Tea Taste Quality Using Miniature NIR Spectroscopy and Electronic Tongue Sensors. *Biosensors* **2023**, *13*, 92. https://doi.org/10.3390/bios13010092

Received: 13 November 2022
Revised: 29 December 2022
Accepted: 4 January 2023
Published: 5 January 2023

1. Introduction

Black tea is the most widely distributed and consumed tea product worldwide and is one of the most important fermented beverages in the world [1]. According to the different processing techniques and shape characteristics, black tea can be divided into Congou black tea, Souchong black tea, and Broken black tea. Among them, Congou black tea is a famous tea product in China, its shape is intact, and the branches are tight and thin, which are processed through four processes, namely withering and losing water in fresh leaves, kneading and forming, fermentation and drying [2]. Compared to Congou black tea, broken black tea is mainly processed with the fresh leaves of large-leaf species, and the kneading process is replaced by kneading and cutting, resulting in a more fragmented or granular shape, which is more conducive to the rapid leaching of nutrients. The main flavor-presenting substances in black tea are tea polyphenols, catechins, free amino acids, caffeine, soluble sugars, inorganic salts, and mineral elements, which are considered to be the main sources of astringency, bitterness, sweetness, and saltiness in black tea, and the composition and proportions of the above active ingredients determine the diversity of the tea's taste [3,4]. To guide consumers in making the right purchasing decisions, prevent fraud, and avoid financial losses, a quick and effective evaluation of the quality of different grades and flavors of black tea has become a pressing issue.

At present, the traditional methods for analyzing and evaluating the quality of tea and its products (i.e., sensory quality evaluation and physical and chemical quality component testing) have been in use for decades [5]. Although the combination of these two types of evaluation methods can accurately determine the content of the intrinsic quality components of tea leaves, effectively assess the quality of tea leaves, and determine the grade of tea leaves, they still have their shortcomings [6]. The sensory quality assessment method is generally used by a professionally trained assessor to determine the quality of a tea sample by scoring it according to its appearance (tenderness, shape, color, etc.) and intrinsic qualities (liquor color, aroma, taste, infused leaf, etc.) combined with weighting factors. This method is susceptible to subjective factors and the objectivity of the results obtained is not sufficient. The physical and chemical composition method generally requires the use of various sophisticated instruments such as liquid chromatography [7,8], gas chromatographs [9], gas chromatograph–mass spectrometers [10,11], liquid chromatograph–mass spectrometers [12,13], and spectrophotometers [14], which are more expensive and have complex sample pre-treatment processes as well as use toxic and dangerous chemicals. Although the test results are accurate and objective, the method has the drawbacks of cumbersome operation, sample destruction, time-consuming and costly testing, and cannot meet the requirements of the rapid multi-component quantitative testing of product quality and online monitoring of characteristic attributes (grade, origin, etc.) during the distribution and processing of tea.

Obtaining objective, fast, and accurate information on the taste and quality characteristics of tea is the ideal method for the digital control of tea. Currently, the main rapid characterization methods widely used in the field of quantitative analysis and qualitative identification of tea taste quality are near-infrared (NIR) spectroscopy and electronic tongue (ET) sensing techniques. NIR spectra are vibrational spectra of molecular leaps of hydrogen-containing groups (bonds such as O—H, C—H, and S—H) in the non-visible region in the wavelength range 780–2526 nm [15]. The technique is more commonly used in the rapid testing of nutrient content and geographical traceability of tea [16]. In recent years, with the development of small and compact portable NIR devices, which minimize the size and manufacturing costs of equipment, the application scenarios of NIR have been expanded, showing good potential for development. The ET is a category of bionic sensing technology that utilizes a specific array of taste sensors to simulate the human tongue for signal perception and pattern recognition of liquid samples [17]. The technique employs suitable multivariate statistical analysis approaches for data processing of the obtained ET response signals to obtain the taste output results of the multi-sensor array, which objectively and rapidly reflects the taste characteristics of the samples and enables fingerprint identification and classification of the test samples [18]. The method has been studied and applied in wine [19], honey [20], tea [17], and other food industries [21]. Currently, most of the published studies evaluate tea quality with a single portable NIR technique or bionic sensing technology [22,23]. Few studies have fused micro-NIR spectroscopy and ET sensor data to more comprehensively characterize black tea quality and grade.

Thus, the main work of this study is organized as follows: (1) Spectral data and ET response signals from 700 black tea samples of seven grades were collected using a home-made portable NIR spectroscopy system and a commercial ET system, and raw data (i.e., low level) fusion was performed; (2) Four feature selection methods, namely grey wolf optimization (GWO), ant colony optimization (ACO), particle swarm optimization (PSO), and non-dominated sorting genetic algorithm II (NSGA−II), were used to extract features (i.e., mid-level data fusion) on the raw fused data; (3) Support vector machine (SVM), extreme learning machine (ELM) and K-nearest neighbor (KNN) algorithm were employed to construct classification models to determine the quality grade of black tea. The experimental procedure for this study is shown in Figure 1.

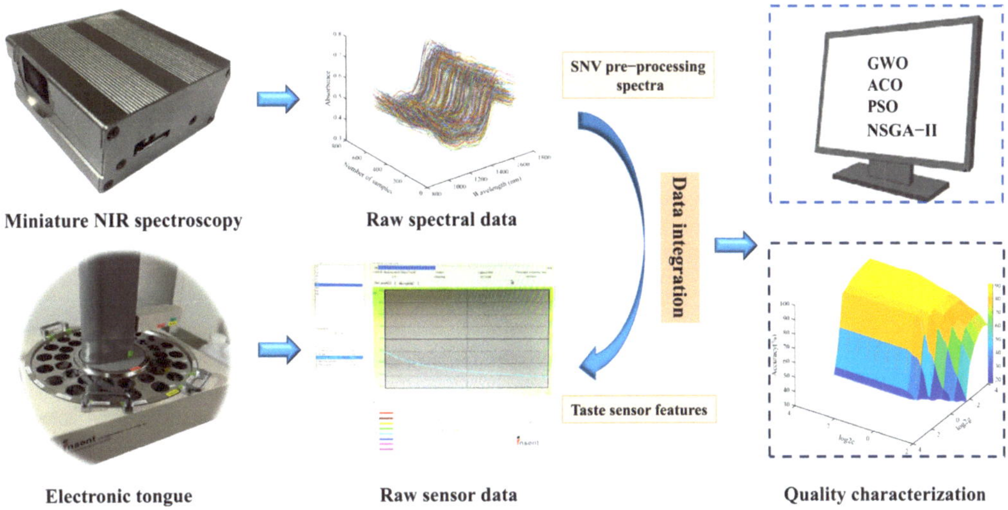

Figure 1. Description of the experimental flow chart.

2. Materials and Methods

2.1. Sample Collection

The experimental samples were seven standard grades of Congou Dianhong black tea products, provided by Yunnan Dian Hong Group Co. (Lincang, China). The contents of the main quality components and the organoleptic evaluation results of the samples were obtained through laboratory testing and sensory evaluation panel scoring by the National Standard Method of the People's Republic of China, respectively, to ensure their quality. The total number of black tea samples was 700, with 100 samples of each grade, in descending order of T, C1, C2, C3, C4, C5, and C6. The grading of black tea samples was mainly determined by their appearance, color, aroma, and flavor when brewed. The tea samples were picked in the spring and autumn of 2019 and 2020. Usually, tea leaves were picked in different seasons to produce a primary processing tea, and the different primary processing teas need to be blended to stabilize the quality of the finished tea product; based on this, grading was carried out. Black tea samples were processed through the processes of withering, twisting, fermentation, and drying of fresh leaves. The samples were dried utilizing a dryer at 90 °C for 30 min. The samples were dried to a moisture content of around 7%. Before data collection, samples were stored in kraft aluminum foil bags and kept in a thermostatic desiccator for three months. Additionally, the single tea sample was tightly knotted, dark brown, and oily, with a uniform appearance. Each tea sample weighed 100 g and was used for multi-sensor signal acquisition and analysis of the data.

2.2. Miniature NIR Spectroscopy and Data Pre-Processing

The miniature NIR spectrometer (Figure 1) developed by our group was equipped with a Bluetooth module and could be connected to a Huawei P50 smartphone for spectral data acquisition on the mobile phone. The NIR device had a spectral acquisition range of 900–1700 nm, a resolution of 10 nm, and weighs only 80 g. During the spectral acquisition, each sample was placed into a quartz sample cup for scanning. After ensuring that the bottom of the sample cup was completely covered by the tea sample, the NIRS information of the sample was collected via diffuse reflection mode. A sample spectrum was acquired every 120° of rotation and the average spectral value of the three scans was used as the subsequent spectral data for modeling purposes.

As the raw spectra acquired by the NIR instrument are susceptible to the physical properties of the sample, background information, and noise interference, it is necessary to pre-process the raw spectra to obtain high−quality spectral features. The effect of spectral pre-treatment methods on the subsequent modeling performance have been investigated in several published studies by our group [16]. The results of several papers showed that the SNV spectral pre−processing method was an effective way to eliminate the effects of solid particles, scattering, and light range variability on the NIRS spectrum. Therefore, the spectral information from the SNV pre−processing was used as feature data for subsequent data analysis in this study. The original NIR spectra of the samples and the SNV pre−treatment spectra are shown in Figure 2. As can be seen in Figure 2, SNV filters out the noise at the beginning and end of the spectral interval, and the SNV-processed spectral curve is smoother and of better quality.

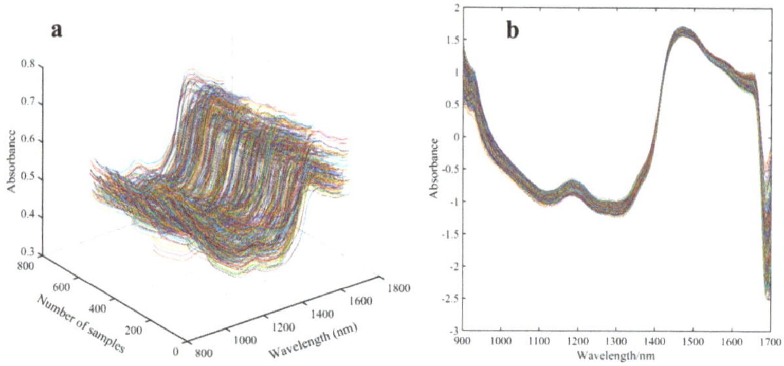

Figure 2. Raw and SNV pre−treatment spectra of samples. (**a**) Raw spectra; (**b**) SNV pre−processing spectra.

2.3. Electronic Tongue Signal Acquisition

The electronic tongue system, model SA402B, was supplied by Insent Intelligent Sensor Technology of Japan and was equipped with an array of six different taste sensors (i.e., CAO, COO, CTO, AE1, AAE, and GL1). The above sensors could perceive nine taste characteristics: sour, bitter, astringent, fresh, salty, sweet, astringent aftertaste, bitter aftertaste, and richness. Before data collection by the ET sensor system, samples were pre-treated in the following sequence: firstly, 3.00 g of tea samples were placed in standard evaluation cups and bowls and brewed for 5 min with 150 mL of boiling distilled water. The tea extract was then filtered through a triple-layer filter cloth and 35 mL of the tea extract was placed into an ET sample cup and cooled at room temperature for the acquisition of ET response signals. The system data acquisition program was set up as follows: 90 s for cleaning the positive and negative electrodes, 120 s for both washes of the reference solution (30 mM KCl and 0.3 mM tartaric acid), and 30 s for sensor data acquisition. The instrument collected the taste potential signal from the sample solution and output it as the nine taste characteristics described above. In the experiment, each sample was measured four times and the average of the four measurements was used as a reference value for data analysis. The working temperature of the test was maintained at 28 °C in circulating water.

2.4. Feature Selection Strategy

To extract feature information from a large number of feature variables related to the exclusive properties of the target substance to be sensed, swarm intelligence optimization algorithms, i.e., PSO, ACO, NSGA−II, and GWO feature variable selection algorithms were introduced to screen ET sensor features and spectra for valid features.

PSO was originally applied as a feature screening method to stimulate social behavior in flocks of birds, fish, and other groups [24]. Considering each bird as each solution of the optimization algorithm in the target space, the population of N particles evolved with each iteration and moved towards the optimal solution of the problem according to the principles of the optimization method, with the particles moving through the path optimization of the previous optimal position and the global optimal position to explore the optimal solution for the whole population [25,26]. The evaluation criterion of the algorithm was determined by the optimal cost parameter of the objective function, the value of which was determined in several iterations of the algorithm [27].

ACO was a swarm optimization algorithm based on the naturally evolving foraging behavior of ant colonies [28]. Scientists studying ants foraging for food discovered that they communicated with each other by spreading pheromones [18]. The algorithm was based on the information feedback mechanism of the ant colony to find the shortest path, enabling intelligent search, parameter optimization, and other functions [29].

NSGA−II was a multi-objective genetic algorithm, proposed by Deb et al. [28]. In the structure of the method, in addition to crossover, genetic operators, and mutation, two multi-objective operators (non-dominated sorting and crowding distance) were utilized. The basic idea of NSGA-II was to hierarchically rank populations by the non-dominant sorting of populations, calculate the crowding distances of individuals to maintain population diversity, and obtain an approximate solution when the termination condition was reached [30].

GWO was a population-based optimization algorithm that simulated the leadership hierarchy and hunting strategy of the natural grey wolf [31]. The algorithm was based on the decision making and management of the alpha wolf leading the pack through the process of tracking, rounding up, and attacking the prey, ultimately achieving the global optimal solution for the capture of the prey [32].

2.5. Modeling Algorithms

The classification models of linear KNN as well as non-linear ELM and SVM were constructed based on the variable information selected by the above methods and their model performance was compared for merit to explore the best classification model for the sample rank.

The KNN algorithm was based on Euclidean distance to explore similar samples and discriminate between different grades of tea samples, and the performance of this classifier depended heavily on the $K-$value and Euclidean distance chosen [33]. The $K-$value was chosen concerning the minimum prediction error of the best result obtained by the classifier [34]. The method was relatively simple and was considered to be one of the fastest machine learning methods to execute on large datasets with uniformly distributed feature spaces [35].

ELM was a machine learning algorithm for multiple classification and regression based on a single implicit layer feed-forward neural network [36]. The discriminator randomly assigned weighting coefficients connecting the input layer and the implicit nodes. The main optimization of the algorithm was described as follows: (1) determine the number of neurons in the hidden layer and randomly setting the input weights of the nodes in the hidden layer; (2) select the activation function of the neurons in the hidden layer (HL) and calculate the output matrix of the neurons in the HL. Based on the highest recognition rate of the prediction set, the optimal number of neurons in the HL is determined [37].

The SVM method was a common multi-classifier employed in data analysis [38]. The algorithm was based on the principle of structural risk minimization and attempted to improve generalization and reduce expected risk [39]. The SVM discriminator used the radial basis function (RBF) as the kernel function and obtained good predictions by optimizing two parameters (i.e., the penalty parameter c and the kernel parameter g) [40]. The specific steps of the method were outlined as follows: (1) the leave−one−out method of cross-validation was employed to optimize the core parameters (c and g); (2) the best parameter pair (c and g) using the grid search method was determined; (3) the best SVM classification model was built based on the highest output of the correct classification rate (CCR) in the prediction set.

2.6. Model Evaluation

In this study, the CCR of the prediction set samples was used as the evaluation criterion for the performance of the model; the higher the CCR value, the higher the prediction accuracy and the better the generalization ability of the model built. All the algorithms for feature selection and qualitative analysis model construction in this work were written by our group and implemented in MATLAB R2020b software (MATLAB Inc., Natick, MA, USA) under Windows 8.1.

3. Results and Discussion

3.1. Sample Set Division and Principal Component Analysis

The Kennard–Stone sample set partitioning algorithm was introduced to obtain the number of the calibration set and prediction set samples of 467 and 233 in sequence with a partitioning ratio of 2:1. The three−dimensional scatter space distribution of the sample set of Dianhong tea samples based on different feature data is shown in Figure 3. As can be seen from Figure 3, the spatial distribution of the single taste features (ET taste values or NIR spectrum) of the Dianhong samples is more discrete from the sample set of the fused data features, with the distribution of the calibration set samples covering the distribution of the prediction set samples. This result indicated that the distribution of the calibration and prediction set samples was appropriate.

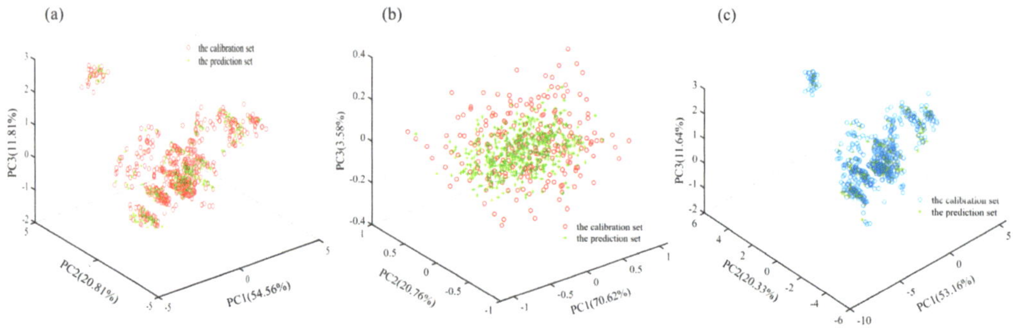

Figure 3. Distributions of Dianhong tea samples from the calibration set and the prediction set in the three−dimensional principal components space. (**a**) ET taste features; (**b**) spectral features; (**c**) data fusion.

Figure 4 shows the results of the three-dimensional principal component analysis (PCA) distribution of the Dianhong tea samples. The distribution of principal component (PC) scores for single taste characteristics (ET taste values and spectra) and fusion data for the seven classes (T, C1, C2, C3, C4, C5, and C6) of samples showed a high degree of overlap in three-dimensional space between the different quality classes, and it was not simple to distinguish the different classes of samples effectively based on both single and fusion feature data, and there was an urgent need to introduce suitable chemometric methods to achieve effective identification of the quality classes of the samples for testing.

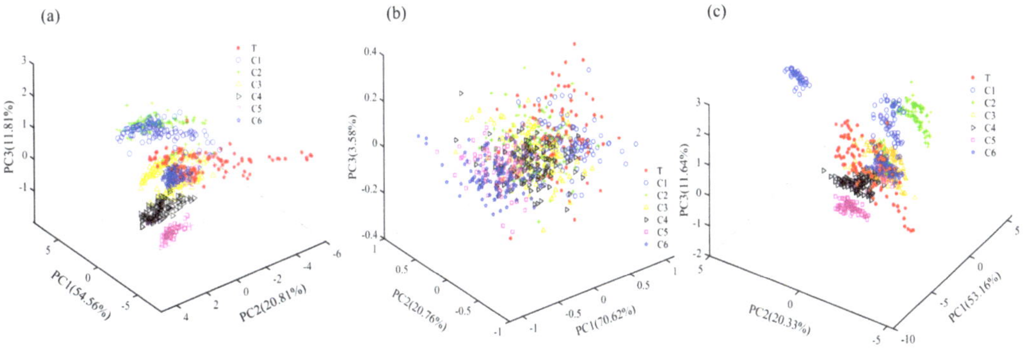

Figure 4. Three−dimensional PC distribution of seven grades of Dianhong tea samples. (**a**) ET taste features; (**b**) spectral features; (**c**) data fusion.

3.2. Selection of Taste Characteristic Variables

Four feature selection methods, namely ACO, PSO, GWO, and NSGA−II, were used to feature the low−level fused data from both sensors (ET and NIR spectra). From the convergence curve of feature selection in Figure 5, it can be seen that the above feature variable selection algorithm eventually converged after several iterations to obtain the smallest objective function value as the optimal solution.

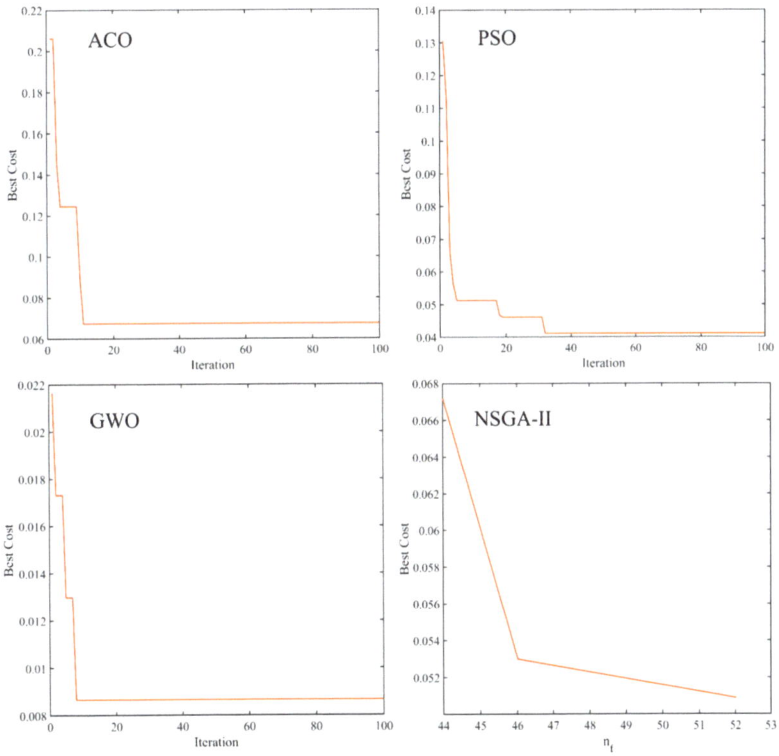

Figure 5. Convergence curves of feature variable selection for different swarm intelligence algorithms.

The results of the wavelength statistics for the above feature variable selection method over 100 iterations are shown in Table 1. As can be seen from Table 1, the number of sample multi-sensor feature variables extracted by the ACO, PSO, GWO, and NSGA−II algorithms were 12, 15, 83, and 52 in respective order, and the number of variables selected as a proportion of the total number of variables was 5.06% (12/237), 6.33% (15/237), 35.02% (83/237), and 21.94% (52/237). The spectral bands chosen for the above methods were mainly in the long-wave region, and the extracted ET response features all contained astringent aftertaste and umami. In summary, by extracting features from the multi-sensor fusion information of the samples, the computational efficiency and complexity of the subsequent modeling could be further simplified, which was more helpful for the construction of high-quality models.

Table 1. Statistical results of different feature variable selection methods within 100 iterations.

Method	Number of Variables	Best Cost	Selected Variables
ACO	12	0.0665	957.03 nm, 977.51 nm, 1004.20 nm, 1024.41 nm, 1286.39 nm,1332.25 nm, 1485.71 nm, 1492.12 nm, 1544.94 nm, 1574.00 nm, astringent aftertaste, umami
PSO	15	0.0411	985.16 nm, 1013.06 nm, 1063.20 nm, 1066.93 nm, 1164.82 nm, 1176.84 nm, 1201.94 nm, 1210.26 nm, 1395.17 nm, 1599.64 nm, 1634.02 nm, bitter aftertaste, astringent aftertaste, umami, saltiness
GWO	83	0.0084	953.17 nm, 964.72 nm, 973.68 nm, 981.33 nm, 985.16 nm, 996.60 nm, 1020.63 nm, 1024.41 nm, 1039.48 nm, 1051.98 nm, 1059.46 nm, 1070.65 nm, 1074.38 nm, 1085.51 nm, 1108.90 nm, 1146.70 nm, 1150.33 nm, 1161.21 nm, 1180.44 nm, 1198.37 nm, 1206.70 nm, 1224.46 nm, 1238.60 nm, 1257.35 nm, 1264.34 nm, 1274.81 nm, 1278.29 nm, 1282.92 nm, 1293.32 nm, 1318.57 nm, 1321.99 nm, 1325.42 nm, 1332.25 nm, 1335.66 nm, 1342.46 nm, 1353.77 nm, 1367.27 nm, 1373.99 nm, 1380.70 nm, 1387.39 nm, 1395.17 nm, 1398.50 nm, 1408.47 nm, 1418.39 nm, 1426.08 nm, 1435.93 nm, 1442.48 nm, 1445.75 nm, 1449.02 nm, 1452.27 nm, 1458.78 nm, 1463.11 nm, 1469.59 nm, 1476.05 nm, 1479.27 nm, 1485.71 nm, 1496.39 nm, 1499.59 nm, 1505.97 nm, 1512.33 nm, 1515.51 nm, 1525.01 nm, 1551.20 nm, 1554.32 nm, 1557.44 nm, 1560.56 nm, 1567.81 nm, 1570.91 nm, 1580.18 nm, 1583.26 nm, 1596.58 nm, 1599.64 nm, 1602.69 nm, 1614.87 nm, 1631.01 nm, 1634.02 nm, 1640.03 nm, sourness, astringency, bitter aftertaste, astringent aftertaste, umami, richness
NSGA−II	52	0.0509	960.88 nm, 969.84 nm, 992.79 nm, 996.60 nm, 1024.41 nm, 1035.71 nm, 1070.65 nm, 1074.38 nm, 1078.09 nm, 1108.90 nm, 1112.58 nm, 1116.25 nm, 1128.47 nm, 1132.12 nm, 1135.77 nm, 1150.33 nm, 1169.64 nm, 1194.79 nm, 1213.82 nm, 1235.07 nm, 1238.60 nm, 1246.82 nm, 1250.33 nm, 1325.42 nm, 1339.06 nm, 1353.77 nm, 1363.90 nm, 1391.84 nm, 1418.39 nm, 1429.37 nm, 1432.65 nm, 1435.93 nm, 1442.48 nm, 1445.75 nm, 1449.02 nm, 1472.82 nm, 1488.92 nm, 1541.81 nm, 1554.32 nm, 1570.91 nm, 1589.42 nm, 1596.58 nm, 1605.74 nm, 1611.83 nm, 1617.91 nm, 1620.94 nm, 1623.97 nm, sourness, bitter aftertaste, astringent aftertaste, umami, sweetness

3.3. Results of The Optimal Models

The results of the optimization models built based on ET sensors and spectral features combined with different chemometrics are shown in Table 2. The statistics in Table 2 showed that the order of CCRs for models based on different data using the same modeling approach was roughly low−level fusion data > ET data > NIR data. The CCR of the KNN model based on ET data only was higher than that of the KNN model with fused data. The reason may be that there were more variables in the fused data and the relationships between them were complex, and KNN as a class of linear algorithms was not as advantageous in solving non-linear complex problems. In terms of the performance of models built from the same data source combined with different classification algorithms, the SVM model has the highest CCR. The highest CCR of 92.27% was obtained from the SVM model based on low−level fused data. It could be seen that the modeling performance utilizing fused data outperformed the results of models constructed from single sensor features.

Table 2. Results of the optimal models based on ET sensors and spectral features combined with different chemometrics.

Data	Model	Parameters	CCR/%			
			Calibration Set		Prediction Set	
ET	ELM	nn [a] = 45	358/467	76.66	178/233	76.39
	KNN	PCs = 4, K = 1	408/467	87.37	201/233	86.27
	SVM	c = 16, g = 5.66	422/467	90.36	210/233	90.13
NIR	ELM	nn [a] = 38	328/467	70.24	150/233	64.38
	KNN	PCs = 3 K = 3	333/467	71.31	152/233	65.24
	SVM	c = 16, g = 0.062	425/467	91.01	198/233	84.98
Low−level Fused data	ELM	nn [a] = 97	376/467	80.51	185/233	79.40
	KNN	PCs = 6, K = 5	370/467	79.23	182/233	78.11
	SVM	c = 16, g = 0.062	455/467	97.43	215/233	92.27

[a] Number of hidden layer neurons.

The feature fusion data obtained using different variable selection algorithms are the characteristic variables and the prediction results of the models built by combining different machine learning algorithms, as shown in Table 3. The results in Table 3 showed that the accuracy of the prediction set for all models was above 80%. The order of merit of the model performance for the same variable selection method was SVM > KNN > ELM. The CCRs of the ACO−ELM, ACO−KNN, and ACO−SVM models were the highest when analyzed in terms of the model effects of the same modeling approach. It can be seen that the ACO algorithm effectively extracted the feature variables related to the quality grade of Dianhong black tea samples. With fewer variables, the model achieved higher accuracy. The optimal classification model that was built was ACO−SVM with a CCR of 93.56% and only 12 variables were extracted. The ACO−SVM model was more robust than models built from fused data at a low level. To explain in detail the discriminative correctness of the prediction set samples in the ACO−SVM model, the confusion matrix results of the prediction sample distribution are presented in Figure 6. The results of the sample distribution in Figure 6 showed that the CCRs for the seven grades (T, C1, C2, C3, C4, C5, and C6) in the prediction class were 100%, 97.1%, 90.9%, 90.9%, 88.2%, 87.9%, and 100%, respectively. The CCRs for the corresponding seven levels in the true class were 91.7%, 97.1%, 96.8%, 93.8%, 96.8%, 85.3%, and 94.3%, respectively. The plotted confusion matrix gave a visual indication of the classification of the samples. The ACO-SVM discrimination model based on fused features was more effective in evaluating the quality of seven categories of tea samples.

Table 3. Results of optimization models based on fused data and different chemometric methods.

Model	Parameters	CCR/%			
		Calibration Set		Prediction Set	
GWO−ELM	nn [a] = 73	397/467	85.01	196/233	84.12
GWO−KNN	PCs = 5, K = 3	408/467	87.37	203/233	87.12
GWO−SVM	c = 16, g = 0.25	450/467	96.36	216/233	92.70
ACO−ELM	nn [a] = 83	398/467	85.22	197/233	84.55
ACO−KNN	PCs = 9, K = 7	421/467	90.15	210/233	90.13
ACO−SVM	c = 16, g = 0.70	448/467	95.93	218/233	93.56
PSO−ELM	nn [a] = 97	386/467	82.66	189/233	81.12
PSO−KNN	PCs = 8, K = 7	421/467	90.15	208/233	89.27
PSO−SVM	c = 11.31, g = 2.83	440/467	94.22	217/233	93.13
NSGA−II−ELM	nn [a] = 73	394/467	84.37	195/233	83.69
NSGA−II−KNN	PCs = 6, K = 5	412/467	88.22	205/233	87.98
NSGA−II−SVM	c = 16, g = 0.35	455/467	97.43	216/233	92.70

[a] Number of hidden layer neurons.

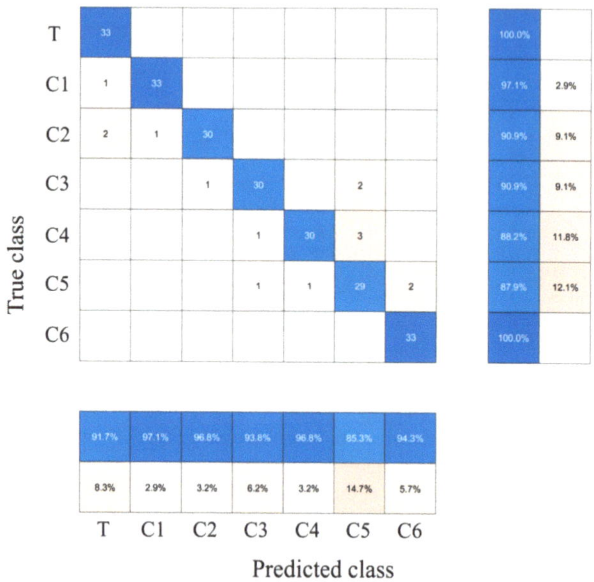

Figure 6. Confusion matrix of the sample distribution of the ACO−SVM model in the prediction process.

3.4. Discussion of the Optimal Models

The characteristic variable features in this study were based on fused data (Miniature NIR data + ET sensor data), and the published papers were all based on a single variable for feature selection [18,41]. The fused data covered a more comprehensive range of information, for which the selection of characteristic variables was beneficial in obtaining valid variables. The results of the selection of feature variables in Table 1 showed that the ACO method obtained the 12 best variables. The number of variables obtained was less than the number of features obtained based on a single spectral information. In published studies based on single spectral data, the performance of the best classification model was also lower than that of the existing ACO−SVM model. Due to the small amount of information from the ET sensor, it was difficult to obtain useful information effectively. Therefore, the stability of the constructed model based on the ET data was insufficient. In terms of the performance of the modeling algorithm, the SVM model had a higher CCR. It had been shown through previous studies that SVM exhibited excellent classification ability in solving non-linear problems [24,42]. In this research, satisfactory results were obtained with the ACO−SVM model.

In the future, the importance of developing small and precise portable NIR equipment to minimize equipment size and manufacturing costs and achieve process control of tea processing quality is an important research direction. In addition, the construction of a cloud-based control system for big data will be an inevitable trend in the development of NIR technology in the field of tea analysis applications.

4. Conclusions

NIR spectroscopy and ET sensors can capture the flavor characteristics of tea and thus assess the flavor quality of tea. In this study, a rapid evaluation method for the taste quality (ET, NIR, and fusion data) of Congou black tea was proposed. The effects of different feature wavelength selection methods (ACO, PSO, GWO, and NSGA−II) and intelligent classification algorithms (KNN, ELM, and SVM) on modeling ET features, NIR features, and multi-sensor feature fusion data of Dianhong black tea samples were explored to find the best assessment parameters and recognition models for black tea grade quality. The experimental results showed that the fusion data (ET + NIR) were filtered for features using the feature variable selection algorithms, and the effective fusion recognition models for the quality of Dianhong black tea were established in combination with the classification methods. The discriminative accuracy of the ACO−SVM model based on fused feature vectors was higher at 93.56% compared to the predictive performance of the single sensor data model. It can be seen that the effective fusion of feature data can reflect the intrinsic properties of the samples to be tested more comprehensively, and the fusion based on ET and spectra has good prospects for evaluating the quality of Dianhong black tea.

Author Contributions: Conceptualization, G.R. and Z.Z.; methodology, G.R.; software, X.Z.; validation, R.W., L.Y. and W.H.; formal analysis, R.W.; investigation, L.Y.; resources, Z.Z.; data curation, G.R.; writing—original draft preparation, G.R.; writing—review and editing, G.R.; visualization, X.Z.; supervision, R.W.; project administration, Z.Z.; funding acquisition, G.R. and Z.Z. All authors have read and agreed to the published version of the manuscript.

Funding: This study was funded by the Open Fund of State Key Laboratory of Tea Plant Biology and Utilization, Anhui Agricultural University (Project No. SKLTOF20220127), the National Key Research and Development Program of China (Project No.2017YFD0400800), and the High-Level Talent Fund Project of Huainan Normal University (Project No. 621222-BSKYQDJ).

Institutional Review Board Statement: Not applicable.

Informed Consent Statement: Not applicable.

Data Availability Statement: Not applicable.

Conflicts of Interest: The authors declare no conflict of interest.

References

1. Li, L.; Cui, Q.; Li, M.; Li, T.; Cao, S.; Dong, S.; Wang, Y.; Dai, Q.; Ning, J. Rapid detection of multiple colorant adulteration in Keemun black tea based on hemp spherical AgNPs-SERS. *Food Chem.* **2023**, *398*, 133841. [CrossRef] [PubMed]
2. Ren, G.; Fan, Q.; He, X.; Li, W.; Tang, X. Applicability of multifunctional preprocessing device for simultaneous estimation of spreading of green tea, withering of black tea and shaking of oolong tea. *J. Sci. Food Agric.* **2020**, *100*, 560–569. [CrossRef] [PubMed]
3. Han, Z.; Wen, M.; Zhang, H.; Zhang, L.; Wan, X.; Ho, C.-T. LC-MS based metabolomics and sensory evaluation reveal the critical compounds of different grades of Huangshan Maofeng green tea. *Food Chem.* **2022**, *374*, 131796. [CrossRef] [PubMed]
4. Ren, Y.; Hou, Y.; Granato, D.; Zha, M.; Xu, W.; Zhang, L. Metabolomics, sensory evaluation, and enzymatic hydrolysis reveal the effect of storage on the critical astringency-active components of crude Pu-erh tea. *J. Food Compost. Anal.* **2022**, *107*, 104387. [CrossRef]
5. Wang, Y.; Cui, Q.; Jin, S.; Zhuo, C.; Luo, Y.; Yu, Y.; Ning, J.; Zhang, Z. Tea Analyzer: A low-cost and portable tool for quality quantification of postharvest fresh tea leaves. *LWT-Food Sci. Technol.* **2022**, *159*, 113248. [CrossRef]
6. Wang, Z.; Gan, S.; Sun, W.; Chen, Z. Quality characteristics of Oolong tea products in different regions and the contribution of thirteen phytochemical components to its taste. *Horticultura* **2022**, *8*, 278. [CrossRef]
7. Li, Q.; Jin, Y.; Jiang, R.; Xu, Y.; Zhang, Y.; Luo, Y.; Huang, J.; Wang, K.; Liu, Z. Dynamic changes in the metabolite profile and taste characteristics of Fu brick tea during the manufacturing process. *Food Chem.* **2021**, *344*, 128576. [CrossRef]
8. Shen, S.; Huang, J.; Li, T.; Wei, Y.; Xu, S.; Wang, Y.; Ning, J. Untargeted and targeted metabolomics reveals potential marker compounds of an tea during storage. *LWT-Food Sci. Technol.* **2022**, *154*, 112791. [CrossRef]
9. Yang, Y.; Xie, J.; Chen, J.; Deng, Y.; Shen, S.; Hua, J.; Wang, J.; Zhu, J.; Yuan, H.; Jiang, Y. Characterization of N,O-heterocycles in green tea during the drying process and unraveling the formation mechanism. *Food Control* **2022**, *139*, 109079. [CrossRef]
10. Li, H.; Geng, W.; Zhang, M.; He, Z.; Haruna, S.A.; Ouyang, Q.; Chen, Q. Qualitative and quantitative analysis of volatile metabolites of foodborne pathogens using colorimetric-bionic sensor coupled robust models. *Microchem. J.* **2022**, *177*, 107282. [CrossRef]
11. Barbosa, M.F.; Santos, J.R.B.; Silva, A.N.; Soares, S.F.C.; Araujo, M.C.U. A cheap handheld NIR spectrometric system for automatic determination of methane, ethane, and propane in natural gas and biogas. *Microchem. J.* **2021**, *170*, 106752. [CrossRef]
12. Ito, A.; Yanase, E. Study into the chemical changes of tea leaf polyphenols during japanese black tea processing. *Food Res. Int.* **2022**, *160*, 111731. [CrossRef]
13. Lin, J.; Liu, F.; Zhou, X.; Tu, Z.; Chen, L.; Wang, Y.; Yang, Y.; Wu, X.; Lv, H.; Zhu, H.; et al. Effect of red light on the composition of metabolites in tea leaves during the withering process using untargeted metabolomics. *J. Sci. Food Agric.* **2022**, *102*, 1628–1639. [CrossRef]

14. Luo, X.; Gouda, M.; Perumal, A.B.; Huang, Z.; Lin, L.; Tang, Y.; Sanaeifar, A.; He, Y.; Li, X.; Dong, C. Using surface-enhanced Raman spectroscopy combined with chemometrics for black tea quality assessment during its fermentation process. *Sens. Actuators B Chem.* **2022**, *373*, 132680. [CrossRef]
15. Ren, G.; Wang, Y.; Ning, J.; Zhang, Z. Highly identification of keemun black tea rank based on cognitive spectroscopy: Near infrared spectroscopy combined with feature variable selection. *Spectrochim. Acta A Mol. Biomol. Spectrosc.* **2020**, *230*, 118079. [CrossRef]
16. Ren, G.; Wang, S.; Ning, J.; Xu, R.; Wang, Y.; Xing, Z.; Wan, X.; Zhang, Z. Quantitative analysis and geographical traceability of black tea using Fourier transform near-infrared spectroscopy (FT-NIRS). *Food Res. Int.* **2013**, *53*, 822–826. [CrossRef]
17. Yang, Z.; Miao, N.; Zhang, X.; Li, Q.; Wang, Z.; Li, C.; Sun, X.; Lan, Y. Employment of an electronic tongue combined with deep learning and transfer learning for discriminating the storage time of Pu-erh tea. *Food Control* **2021**, *121*, 107608. [CrossRef]
18. Ren, G.; Li, T.; Wei, Y.; Ning, J.; Zhang, Z. Estimation of Congou black tea quality by an electronic tongue technology combined with multivariate analysis. *Microchem. J.* **2021**, *163*, 105899. [CrossRef]
19. Zaukuu, J.-L.Z.; Soós, J.; Bodor, Z.; Felföldi, J.; Magyar, I.; Kovacs, Z. Authentication of tokaj wine (*Hungaricum*) with the electronic tongue and near infrared spectroscopy. *J. Food Sci.* **2019**, *84*, 3437–3444. [CrossRef]
20. Sobrino-Gregorio, L.; Tanleque-Alberto, F.; Bataller, R.; Soto, J.; Escriche, I. Using an automatic pulse voltammetric electronic tongue to verify the origin of honey from Spain, Honduras, and Mozambique. *J. Sci. Food Agric.* **2020**, *100*, 212–217. [CrossRef]
21. Lu, L.; Hu, Z.; Hu, X.; Han, J.; Zhu, Z.; Tian, S.; Chen, Z. Quantitative approach of multidimensional interactive sensing for rice quality using electronic tongue sensor array based on information entropy. *Sens. Actuators B Chem.* **2021**, *329*, 129254. [CrossRef]
22. Li, L.; Wang, Y.; Jin, S.; Li, M.; Chen, Q.; Ning, J.; Zhang, Z. Evaluation of black tea by using smartphone imaging coupled with micro-near-infrared spectrometer. *Spectrochim. Acta A Mol. Biomol. Spectrosc.* **2021**, *246*, 118991. [CrossRef] [PubMed]
23. Ouyang, Q.; Yang, Y.; Wu, J.; Chen, Q.; Guo, Z.; Li, H. Measurement of total free amino acids content in black tea using electronic tongue technology coupled with chemometrics. *LWT-Food Sci. Technol.* **2020**, *118*, 108768. [CrossRef]
24. Ding, Y.; Yan, Y.; Li, J.; Chen, X.; Jiang, H. Classification of tea quality levels using near-infrared spectroscopy based on CLPSO-SVM. *Foods* **2022**, *11*, 1658. [CrossRef] [PubMed]
25. Liu, T.; Jiang, H.; Chen, Q. Input features and parameters optimization improved the prediction accuracy of support vector regression models based on colorimetric sensor data for detection of aflatoxin B1 in corn. *Microchem. J.* **2022**, *178*, 107407. [CrossRef]
26. Zhao, J.; Tian, G.; Qiu, Y.; Qu, H. Rapid quantification of active pharmaceutical ingredient for sugar-free Yangwei granules in commercial production using FT-NIR spectroscopy based on machine learning techniques. *Spectrochim. Acta A Mol. Biomol. Spectrosc.* **2021**, *245*, 118878. [CrossRef]
27. Jiang, H.; Liu, T.; He, P.; Ding, Y.; Chen, Q. Rapid measurement of fatty acid content during flour storage using a color-sensitive gas sensor array: Comparing the effects of swarm intelligence optimization algorithms on sensor features. *Food Chem.* **2021**, *338*, 127828. [CrossRef]
28. Liu, T.; He, J.; Yao, W.; Jiang, H.; Chen, Q. Determination of aflatoxin B1 value in corn based on Fourier transform near-infrared spectroscopy: Comparison of optimization effect of characteristic wavelengths. *LWT-Food Sci. Technol.* **2022**, *164*, 113657. [CrossRef]
29. Modupalli, N.; Naik, M.; Sunil, C.K.; Natarajan, V. Emerging non-destructive methods for quality and safety monitoring of spices. *Trends Food Sci. Technol.* **2021**, *108*, 133–147. [CrossRef]
30. Cordeiro, F.R.; da Silva-Filho, A.G. Multi-objective optimization applied to unified second level cache memory hierarchy tuning aiming at energy and performance optimization. *Appl. Soft. Comput* **2016**, *49*, 603–610. [CrossRef]
31. Tu, Q.; Chen, X.; Liu, X. Multi-strategy ensemble grey wolf optimizer and its application to feature selection. *Appl. Soft. Comput* **2019**, *76*, 16–30. [CrossRef]
32. Too, J.; Abdullah, A.R. Opposition based competitive grey wolf optimizer for EMG feature selection. *Evol. Intell.* **2021**, *14*, 1691–1705. [CrossRef]
33. Panero, F.D.; Smiderle, O.; Panero, J.S.; Faria, F.S.D.V.; Panero, P.d.S.; Rodriguez, A.F.R. Non-destructive genotyping of cultivars and strains of sesame through NIR spectroscopy and chemometrics. *Biosensors* **2022**, *12*, 69. [CrossRef]
34. Ren, G.; Sun, Y.; Li, M.; Ning, J.; Zhang, Z. Cognitive spectroscopy for evaluating Chinese black tea grades (*Camellia sinensis*): Near-infrared spectroscopy and evolutionary algorithms. *J. Sci. Food Agric.* **2020**, *100*, 3950–3959. [CrossRef]
35. Miao, X.; Miao, Y.; Tao, S.; Liu, D.; Chen, Z.; Wang, J.; Huang, W.; Yu, Y. Classification of rice based on storage time by using near infrared spectroscopy and chemometric methods. *Microchem. J.* **2021**, *171*, 106841. [CrossRef]
36. Dong, C.; An, T.; Yang, M.; Yang, C.; Liu, Z.; Li, Y.; Duan, D.; Fan, S. Quantitative prediction and visual detection of the moisture content of withering leaves in black tea (*Camellia sinensis*) with hyperspectral image. *Infrared Phys. Technol.* **2022**, *123*, 104118. [CrossRef]
37. An, T.; Yu, S.; Huang, W.; Li, G.; Tian, X.; Fan, S.; Dong, C.; Zhao, C. Robustness and accuracy evaluation of moisture prediction model for black tea withering process using hyperspectral imaging. *Spectrochim. Acta A Mol. Biomol. Spectrosc.* **2022**, *269*, 120791. [CrossRef]
38. Jia, H.; Yuan, W.; Ren, Z.; Ning, J.; Xu, Y.-Q.; Wang, Y.; Deng, W.-W. Cost-effective and sensitive indicator-displacement array (IDA) assay for quality monitoring of black tea fermentation. *Food Chem.* **2023**, *403*, 134340. [CrossRef]

39. Bian, X.; Wu, D.; Zhang, K.; Liu, P.; Shi, H.; Tan, X.; Wang, Z. Variational mode decomposition weighted multiscale support vector regression for spectral determination of rapeseed oil and rhizoma alpiniae offcinarum adulterants. *Biosensors* **2022**, *12*, 586. [CrossRef]
40. Wang, Y.; Ren, Z.; Li, M.; Yuan, W.; Zhang, Z.; Ning, J. pH indicator-based sensor array in combination with hyperspectral imaging for intelligent evaluation of withering degree during processing of black tea. *Spectrochim. Acta A Mol. Biomol. Spectrosc.* **2022**, *271*, 120959. [CrossRef]
41. Ren, G.; Zhang, X.; Wu, R.; Zhang, X.; Xie, T.; Zhang, Z. Digital depiction of the quality of Dianhong black tea based on pocket-sized near infrared spectroscopy. *Infrared Phys. Technol.* **2022**, *127*, 104418. [CrossRef]
42. Jia, J.; Zhou, X.; Li, Y.; Wang, M.; Liu, Z.; Dong, C. Establishment of a rapid detection model for the sensory quality and components of Yuezhou Longjing tea using near-infrared spectroscopy. *LWT-Food Sci. Technol.* **2022**, *164*, 113625. [CrossRef]

biosensors

MDPI

Article

A Real-Time Detection Method of Hg^{2+} in Drinking Water via Portable Biosensor: Using a Smartphone as a Low-Cost Micro-Spectrometer to Read the Colorimetric Signals

Yifan Gu [1,2,†], Leizi Jiao [2,†], Fengjing Cao [2], Xinchao Liu [1], Yunhai Zhou [2], Chongshan Yang [2], Zhen Gao [2], Mengjie Zhang [1,2], Peng Lin [1,2], Yuxing Han [3,4,*] and Daming Dong [2,*]

1 College of Electronic Engineering (College of Artificial Intelligence), South China Agricultural University, Guangzhou 510642, China
2 National Research Center of Intelligent Equipment for Agriculture, Beijing Academy of Agriculture and Forestry Sciences, Beijing 100097, China
3 Shenzhen International Graduate School, Tsinghua University, Shenzhen 518055, China
4 RIOS Lab, Tsinghua University, Shenzhen 518055, China
* Correspondence: yuxinghan@sz.tsinghua.edu.cn (Y.H.); damingdong@hotmail.com (D.D.)
† These authors contributed equally to this work.

Abstract: This paper reported a real-time detection strategy for Hg^{2+} inspired by the visible spectrophotometer that used a smartphone as a low-cost micro-spectrometer. In combination with the smartphone's camera and optical accessories, the phone's built-in software can process the received light band image and then read out the spectral data in real time. The sensor was also used to detect gold nanoparticles with an LOD of 0.14 μM, which are widely used in colorimetric biosensors. Ultimately, a gold nanoparticles-glutathione (AuNPs-GSH) conjugate was used as a probe to detect Hg^{2+} in water with an LOD of 1.2 nM and was applied successfully to natural mineral water, pure water, tap water, and river water samples.

Keywords: colorimetric sensor; smartphone; micro-spectrometer; gold nanoparticle; Hg^{2+} detection

Citation: Gu, Y.; Jiao, L.; Cao, F.; Liu, X.; Zhou, Y.; Yang, C.; Gao, Z.; Zhang, M.; Lin, P.; Han, Y.; et al. A Real-Time Detection Method of Hg^{2+} in Drinking Water via Portable Biosensor: Using a Smartphone as a Low-Cost Micro-Spectrometer to Read the Colorimetric Signals. *Biosensors* **2022**, *12*, 1017. https://doi.org/10.3390/bios12111017

Received: 30 September 2022
Accepted: 9 November 2022
Published: 14 November 2022

1. Introduction

The distribution of mercury pollution within the various spheres of the Earth system has been attracting long-term attention because of the substantial toxicity of mercury [1]. Hg^{2+} is one of the most stable inorganic forms of mercury and has characteristics highly hazardous to health at low concentrations [2], such as permanent damage to the human digestive system, urinary system, and nervous system. Therefore, to prevent Hg^{2+} from threatening the health of human beings and ecosystems as much as possible, it is necessary to perform qualitative and quantitative detection of the trace levels of Hg^{2+} in water.

Several well-established methods are currently available for detection of heavy metals in drinking water and in its source, including atomic absorption spectroscopy [3,4], atomic fluorescence spectroscopy [5,6], inductively-coupled plasma mass spectrometry [7,8], and gas chromatography-mass spectrometry (GC-MS) [9]. Many rapid-detection methods have been used to improve detection efficiency, including colorimetric methods based on the nucleic acids, antibodies, nanomaterials, or paper [10,11], fluorescence methods [12,13], electrochemical conduction methods [14,15], and recently-developed surface Raman-enhanced scattering methods [16]. Although these methods have been shown to provide high sensitivity and accuracy using biological and chemical reactions, there are some drawbacks that must still be addressed, such as difficulty in preparing nucleic acid aptamers or antibodies, low ligation efficiency, extended response time, complexity, and short lifetime. In contrast, colorimetric biosensors based on nanotechnology are more attractive because of their simple preparation processes, low cost, and excellent readability [17–19]. Gold nanoparticles

(AuNPs) are favorable for colorimetric biosensing applications due to the high molar extinction coefficient, high specific surface area, and easy functionalization, leading to color changes that are related to their interparticle distances [20–23]. Additionally, the spectrophotometer is widely used in biological colorimetric sensors as an important analytical device. However, the cost of spectrophotometers for non-laboratory-based applications is high, the available software support is severely limited due to the scalability by specific permissions, and the device is strongly dependent on professional and proficient operators with regard to data processing [24]. Additionally, an adequate platform is required to place and power the spectrometer. Therefore, the development of a portable spectrophotometer is essential to reduce detection costs and increase simplicity, and smartphones have gradually emerged as an outstanding option [25–27].

Research trends and innovative approaches in the testing field have begun to focus more on light device weight, simplicity, and effectiveness. Over the past decade, smartphones have become indispensable personal devices for many people and are attracting considerable attention for low-cost, low-power, portable, and high-speed biosensors [28–30]. Among the device features, complementary metal-oxide-semiconductor (CMOS) image sensors on smartphones can convert optical signals into electrical signals [31]. With the continuous advancements in their integrated circuit design and manufacturing technology, smartphone cameras are increasingly improving sensitivity, resolution, and dynamic range with reduced power consumption [32,33]. These cameras have even been applied to smart cars, which are more demanding than smartphones [34]. These advantages enable smartphones to be used as readout tools for portable colorimetric biosensors to measure color and intensity changes effectively.

Based on the studies referenced above, we constructed a colorimetric biosensor system that uses a smartphone miniature spectrometer as a readout tool (Figure 1A) in this work. As a part of evaluating the sensor, we used a monochromatic diode and a gold nanoparticles (AuNPs) solution to evaluate the accuracy and stability of the proposed system. We also used gold nanoparticles-glutathione (AuNPs-GSH) conjugate as a detection probe to test the practical applicability of the proposed sensor (Figure 1B). According to the three classifications of drinking water in China's national standard (GB19298-2014, GB5749-2006) and an essential source of drinking water, we selected natural mineral water, pure water, tap water, and river water as actual samples for the spiking experiment and obtained satisfactory results.

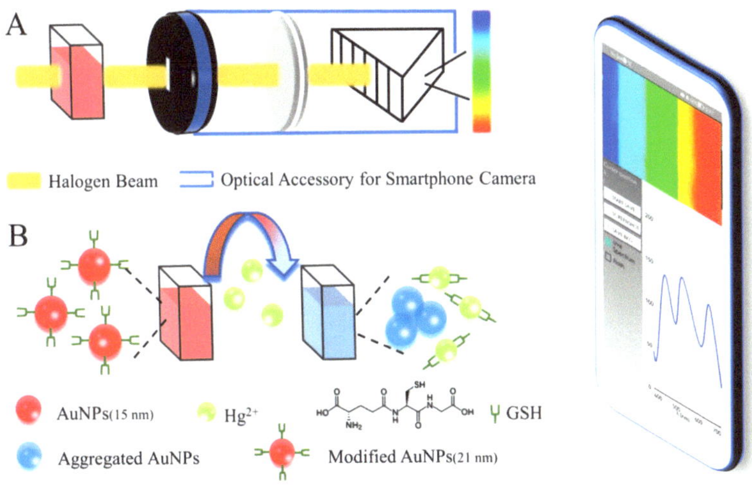

Figure 1. (**A**) Colorimetric biosensor system based on a smartphone; (**B**) schematic illustration of Hg^{2+} detection based on AuNPs-GSH.

2. Experimental Section

2.1. *Materials and Instruments*

The details have been listed in Supporting Information.

2.2. *Preparation of AuNPs and AuNPs-GSH*

Gold nanoparticles (AuNPs) were prepared using the trisodium citrate reduction method established by Frens in 1973 [35]. We added 85.8 mL of ultrapure water to a three-necked round-bottomed flask, followed by 4.2 mL of chloroauric acid solution (1%), and then the solution was heated to 120 °C and stirred under magnetic reflux until boiling. Then, 10 mL of trisodium citrate solution (1%) was added rapidly, and the color of the solution turned red within 1 min. We then continued to boil the solution for 20 min and subsequently cooled it to room temperature naturally to obtain a wine-red AuNPs solution. This solution was then stored at 4 °C for later use within a one-year period.

Then, 30 µL of the GSH solution (25 mg/mL) was added to 6 mL of the AuNPs solution. The mixture was placed on a magnetic stirrer and was stirred at room temperature for 2 h to modify the AuNPs completely using GSH via the Au-S bond. The solution was then centrifuged at high speed (30 min, 12,000 rpm, 4 °C) to remove the supernatant containing unreacted GSH, and deionized water was added to re-suspend the AuNPs-GSH conjugates. This solution was also stored at 4 °C for later use within a two-month period.

2.3. *Evaluation of Sensor Capability for Colorimetric Detection*

First, a commercial miniature optical fiber spectrometer (Avantes) was used to detect four narrow-band monochromatic diode light sources. The results obtained were then used as the actual wavelength bands for the monochromatic diodes. After a fixed distance was set between each diode and the smartphone, the smartphone was set to detect the diodes at 2 s detection intervals. In addition, the band position and intensity data of the single peaks were recorded to verify the accuracy and stability of the smartphone's detection performance.

Further detection and absorbance analyses of AuNPs solutions with various concentrations (1, 5, 10, 25, 50, 75, and 100 µM) were performed to verify the colorimetric detection capability of the smartphone. In this experiment, deionized water was used as the background for the absorbance analysis. For comparison and verification, the background and each concentration of AuNPs solution were automatically detected 20 times by an Avantes micro-optical fiber spectrometer and then the average detected values were exported. Under the same conditions, the smartphone performed 20 detections for the same solution and then obtained the average value. In addition, the 1st, 10th, and 20th of 20 detections for each concentration were selected as three parallel experiments of this concentration.

Limit of detection (LOD) was one of the indicators used to evaluate our detection performance. It was calculated by 3 sb/slope (sb: standard deviation of the background, and the slope in the calibration plot).

2.4. *Practical Application of Sensors to Colorimetric Detection*

2.4.1. PH and Response Time Optimization

The pH of the AuNPs-GSH solution was adjusted to values of 3.5, 4.5, 5.5, 6.5, and 7.5. Then, 5 µM and 10 µM Hg^{2+} solutions were added to each of the AuNPs-GSH solutions with different pH values, and the V_{Hg2+}:$V_{AuNPs\text{-}GSH}$ was 2:1.The spectral data of these solutions were detected at 522 nm and recorded every 3 s using the smartphone, and the detection process lasted for 30 s. Each assay in these experiments was repeated three times. In the analysis, the absorbance data of the solutions at 522 nm were used as indicators of the degree of dispersion or aggregation of the AuNPs.

2.4.2. Detection of Specificity

The AuNPs-GSH solution was mixed with 16 common ions (Ni^{2+}, Cd^{2+}, Cr^{3+}, Hg^{2+}, Pb^{2+}, Mn^{2+}, Cu^{2+}, Ca^{2+}, Mg^{2+}, Zn^{2+}, Fe^{3+}, CrO_4^{2-}, CO_3^{2-}, SO_4^{2-}, CH_3COO^-, and PO_4^{3-} ions, 5 μM) at a volume ratio of 1:2 and were detected under the same experimental conditions using the proposed sensor, and each detection process was repeated three times. Ten metal ions (Ni^{2+}, Cd^{2+}, Cr^{3+}, Pb^{2+}, Mn^{2+}, Cu^{2+}, CrO_4^{2-}, Ca^{2+}, Mg^{2+}, Zn^{2+}, and Fe^{3+} ions, 0.5 μM) were mixed. The mixture solution with/without Hg^{2+} (0.5 μM) was added to the AuNPs-GSH solution and detected with the same detection method as above.

2.4.3. Quantitative Detection

We added Hg^{2+} solutions with different concentrations ($V_{AuNPs\text{-}GSH}$:V_{Hg2+} = 1:2), and then each mixture was incubated for 5 s. To better use the smartphone for detection in this case, the mixtures obtained above were then diluted at a ratio of 1:4 with deionized water. The smartphone's spectral data were detected and recorded from 400 nm to 700 nm. The relative distances between the light source, the cuvette, and the smartphone were all fixed. The total detection time was approximately 300 s. The detection time refers to the time between the dropwise addition of the first Hg^{2+} solution concentration to the AuNPs-GSH solution and the detection of the final Hg^{2+} solution concentration. The experiment was repeated three times. Among the results, the absorbance data acquired at 522 nm were used to provide the index parameter for the Hg^{2+} content, which also represented the degree of aggregation or dispersion of the AuNPs.

3. Results and Discussion

3.1. Smartphone Functions of the Sensor

The colorimetric biosensing system constructed in this work used the smartphone camera as a readout tool. The light band image that was formed by the light beam when dispersed by the optical accessories reached the smartphone's camera. The smartphone then collected the detected spectral data based on the image received (this principle is illustrated in Figure S1, Supplementary Materials). The operational interface of the mobile terminal processing software (GoSpectro app) is shown in Figure 2. The accuracy and stability of the spectrum band data obtained are vital to the detection process. The calibration process can be adjusted using the parameter settings before detection and the calibration light source. After the calibration process is complete, the smartphone software will then match the pixel position and intensity of the received image with the position and intensity of the band. This will enable the formation of the visual and spectral data simultaneously. These data can be read and recorded in real-time. This is true whether the data are acquired for mercury lamps with multiple narrow-band light sources or for halogen lamps with a broader spectrum of wavelengths. We detected mercury lamps to validate that using the above process could obtain available and similar results from different low-configuration smartphones (Table S1 and Figure S2).

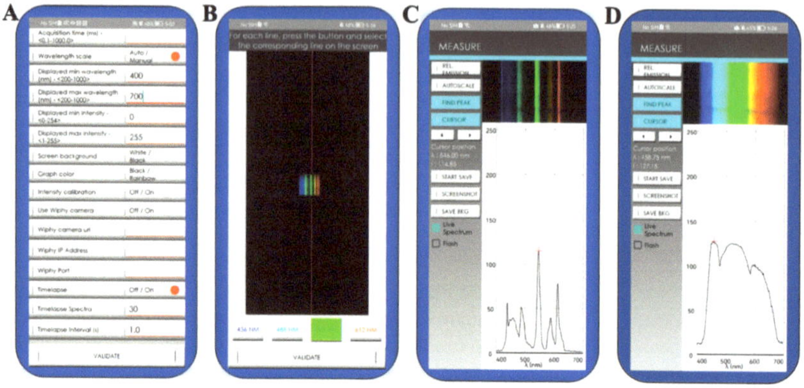

Figure 2. Operating interface of GoSpectro app (Android) on the smartphone. (**A**) Detection parameter settings; (**B**) calibration operation with calibration light source; (**C**) detection of mercury lamp; and (**D**) detection of white light source.

3.2. Evaluation of Sensor Performance in Colorimetric Detection

3.2.1. Accuracy and Stability of Sensor Detection

Using a smartphone as a reading device involves two primary abilities: the wavelength band must be detected accurately, and the light intensity detection must remain relatively stable. This part of the experimental process was verified using four monochromatic diodes with different wavelengths (Figure 3A). The results in Table S2 show that the position of the detection peak of the proposed sensor is basically the same as that of the Avantes spectrometer, and the existing sensor error will not affect the detection significantly (0.5 nm).

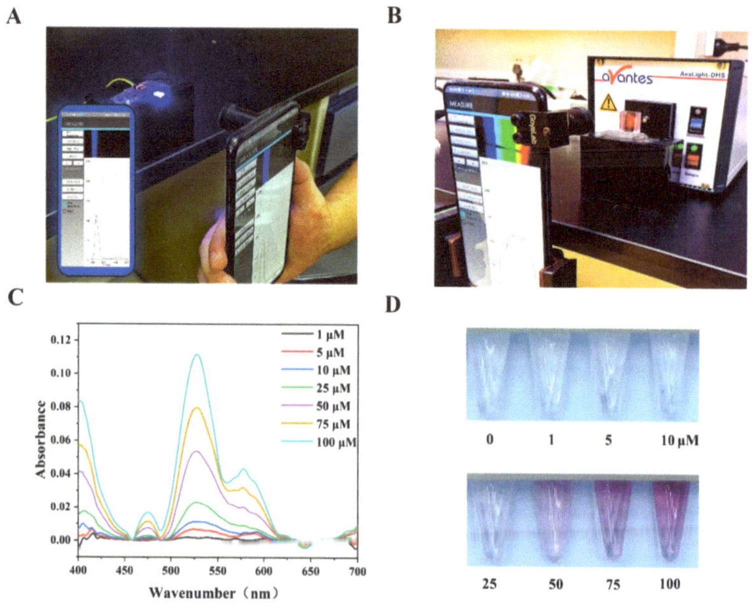

Figure 3. (**A**) Verification of the accuracy and stability of the sensor proposed in this paper via the detection of monochromatic diodes; (**B**) detection of AuNPs solutions via the sensing system; (**C**) absorbance analysis results for AuNPs solutions with different concentrations; (**D**) color changes in the solutions with increasing AuNPs concentrations.

These results also verified that the calibration operation performed before detection was effective. Otherwise, the calibration would cause the image pixels to fail to correspond to the band's position, and the detection peak would have been shifted. Furthermore, the fluctuations in the light intensity data for the detection peak are minimal, thus indicating that the impact of the system error is limited. This illustrates the good repeatability of the detection operation.

The cooperation between the smartphone camera and the optical accessories also indicates that the sensing system can provide an accurate and stable detection performance after a standard calibration operation.

3.2.2. Sensitivity Analysis of Sensor Detection

For colorimetric detection applications, it is imperative that the sensor be highly sensitive to color. It is possible to change the color of the solution under test gradually, from colorless to wine red, by varying the AuNPs concentration. Therefore, we analyzed the detection and absorbance of AuNPs solutions with different concentrations to verify the smartphone's detection performance (Figure 3B). As shown in Figure 3C, the absorbance analysis indicated that the AuNPs appeared to have a prominent absorption peak at approximately 522 nm due to the 15 nm AuNPs. In addition, within the 625–650 nm range, there is a fluorescence effect that results in the appearance of a negative absorbance peak for this band range. These two characteristic peaks showed good repeatability in multiple experiments, and the Avantes spectrometer under the same experimental conditions verified this phenomenon as well.

We performed linear regression analyses on the absorbance data acquired at 522 nm from both the smartphone and the Avantes spectrometer. These analyses evaluated the detection capability of the smartphone. Both detection devices showed excellent linear relationships between the concentration and the absorbance of the AuNPs within the detection range from 1 μM to 100 μM (Figure S3). The correlation coefficients of the two devices were $R^2 = 0.99922$ and $R^2 = 0.99975$, respectively, and their slopes differed by only 0.0002. Additionally, the limit of detection (LOD, 3 sb/slope) for AuNPs of the smartphone reached 0.14 μM (Figure S3A). These results showed that the sensing system provided a detection capability comparable to that of a commercial spectrometer over an extensive range of variations of the same color. It was noted that even when the AuNPs solution appeared to the naked eye as having little to no color, or when the colors of the solutions were so close to each other that it was difficult to differentiate between them (Figure 3D), the smartphone could detect the gradient successfully. This provides an intuitive illustration of the sensitivity and conveys that the smartphone camera is highly suited to the capture of color changes, thus meeting the colorimetric application requirements.

3.3. *Practical Application of the Sensors to Hg^{2+} Detection*

AuNPs are used widely because of their excellent optical properties. We selected the detection performance for Hg^{2+} sensing based on GSH-modified AuNPs to evaluate the practical applicability of the sensor. This allows the sensor to work under more complex color change conditions to ensure that comprehensive performance testing can be obtained during practical applications. The detection process based on use of AuNPs as probes can also reflect the scalability of the sensor in colorimetric detection.

3.3.1. Principle and Characterization of Hg^{2+} Detection by AuNPs-GSH

AuNPs have a high affinity for biological thiols, and when molecules containing thiol groups are added to the AuNPs solution, they will bind rapidly to the surfaces of the AuNPs [36]. When compared with AuNPs and other metal ions, Hg^{2+} has higher thiophilic properties [37]. Based on this principle, several studies have been performed that have led to the aggregation of AuNPs, which results in a color change from red to blue [38–40] or causes aggregated AuNPs to achieve an anti-aggregation effect, which in turn causes the solution color to change from blue to red as the particles are re-dispersed [41–43]. This enables the detection of Hg^{2+} or biothiols using AuNPs. Based on the studies described above, trisodium citrate-reduced AuNPs (citrate-AuNPs) were modified with GSH via Au-S bonds at room temperature to form a stable AuNPs-GSH probe. When a specific volume of the Hg^{2+} solution was added to the AuNPs-GSH solution, the extremely high affinity of -SH toward Hg^{2+} triggered the breakage of Au-S bonds on surfaces of AuNPs, causing GSH to fall from Au surfaces and then form a GSH-Hg-GSH complex with Hg^{2+}, thus destabilizing the AuNPs and causing aggregation and color changing [44]. When the Hg^{2+} concentration increases, the AuNPs solution follows a red-purple-blue color change pattern, which also places higher requirements on the tests of the proposed sensor's detection performance.

To verify the capability of the practical smartphone detection process and the feasibility of the proposed method, we first characterized the AuNPs-GSH solution using the sensing system presented in this paper (Figure 4). The AuNPs-GSH absorption peak is located near 522 nm. The addition of a high Hg^{2+} concentration causes the AuNPs-GSH solution to aggregate and rapidly change from red to blue. We also observed the process of the AuNPs-GSH solution before and after aggregation by transmission electron microscopy (TEM). Figure 4B showed that the prepared AuNPs-GSH had a good dispersion, but when a specific concentration of the Hg^{2+} solution was added to the AuNPs-GSH solution, the AuNPs then aggregated (Figure 4C). The absorption peak in Figure 4A showed a significant shift to the right with the addition of Hg^{2+}, which is related to the red-shift phenomenon caused by the larger overall diameter of the aggregated AuNPs. Further, we used dynamic light scattering (DLS) to characterize the size distribution changes of AuNPs, AuNPs-GSH, and AuNPs-GSH+Hg^{2+}. The results (Figure S4) showed that the average size of AuNPs is about 15 nm. The hydrodynamic size of AuNPs-GSH increased to about 21 nm after the conjugation of GSH on the surface of AuNPs. We also validated the aggregation process using DLS, which revealed that the hydrodynamic size of AuNPs was dramatically enlarged relative to the AuNPs-GSH when Hg^{2+} was introduced. The zeta potential also was tested as well. Compared with AuNPs, the zeta potential of AuNPs-GSH changed from -44.3 to -26.6 mV, which suggested that GSH has been successfully modified on the surfaces of AuNPs, and it changed to -1.56 mV after adding Hg^{2+}. These characterization processes verify that the AuNPs-GSH-based probe has the potential to detect Hg^{2+}, while the accurate detection of the red-shift phenomenon also illustrates the excellent performance of the proposed sensor.

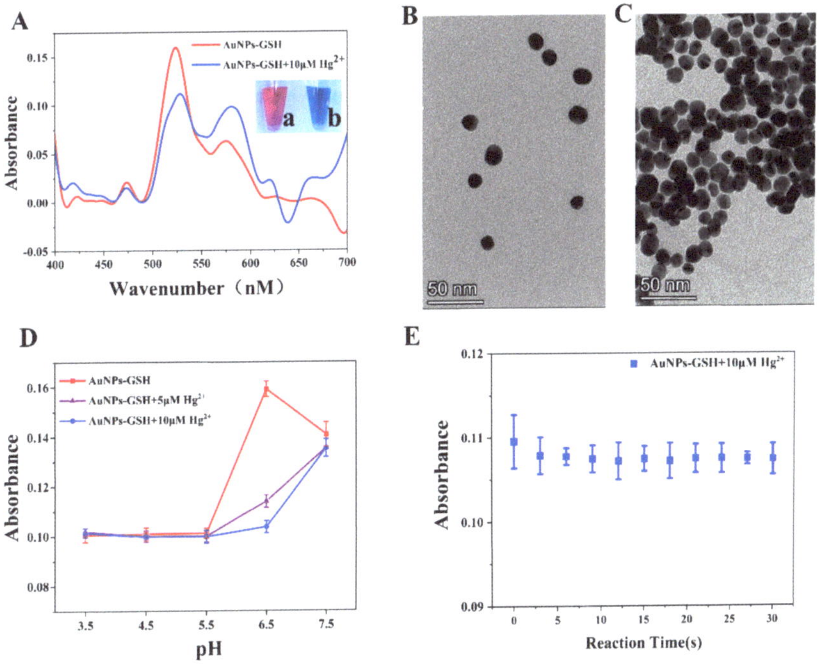

Figure 4. (**A**) Characterization of the absorption spectra of the AuNPs−GSH solution using the proposed sensor (a) without Hg^{2+} and (b) with 10 μM of Hg^{2+}; (**B**,**C**) TEM images of the AuNPs−GSH solution before and after addition of Hg^{2+}; (**D**) absorbance analysis results before and after addition of Hg^{2+} to the AuNPs−GSH solution under different acid−base environmental conditions; (**E**) changes in absorbance over time after addition of Hg^{2+} to the AuNPs−GSH solution.

3.3.2. Optimization of the Experimental Conditions

To remove the influence of certain factors in the experiment of smartphone detection performance and to provide a better reflection for the practical application of the sensor proposed in this paper, we analyzed and optimized the pH value of the solution and the response time to the AuNPs-GSH solution with Hg^{2+}. Figure 4D showed that the AuNPs were extremely unstable in an acidic environment and exhibited low absorbance. From a combination of the absorbance values with the electron microscope images (Figure 4A,D), it can be inferred that the AuNPs were aggregated to a high degree at this time, which also means that even if Hg^{2+} was added to the solution, no significant change would be observed. According to the results of this analysis, the detection should be performed in a weak acid environment. Because the reaction of the solution tends to become stable within approximately 5 s of adding Hg^{2+} (Figure 4E), the incubation time for the probe and target was set at 5 s. In this section, the smartphone detection results demonstrated the device's rapid-detection capability and reflected the aggregation of the AuNPs accurately.

3.3.3. System-Specific Analysis and Quantitative Detection

The specific selection of AuNPs-GSH is essential to guarantee the effective detection of Hg^{2+} by the proposed sensor. In the experiments, we adjusted the volume ratio of the AuNPs-GSH solution to 16 common ions solutions (1:2) appropriately, and the color changed dramatically when Hg^{2+} was dropped into this solution. Additionally, the change in this solution was most obvious in the 5 min following the addition of Hg^{2+} (Figure S5). To better illustrate this feature, we used the smartphone to detect the absorbance at 522 nm after the AuNPs-GSH solution was mixed with common ions (Figure S6), as well as the value of the resulting absorbance change (Figure 5A). When the AuNPs-GSH solution was mixed with deionized water as a control group, it was found that Hg^{2+} produced a very prominent response, whereas the influence of adding other ions to the solution was similar to the control group. To a certain extent, this also showed that interference from the other ions is very limited at the volume ratio of 1:2. Under the same conditions, when we added the mixture solution containing 10 metal ions (excluding Hg^{2+}) to AuNPs-GSH, there was no obvious interference. However, the mixture solution containing Hg^{2+} caused AuNPs aggregation (Figure S7), which showed that our method still had good selectivity when ions competed with each other.

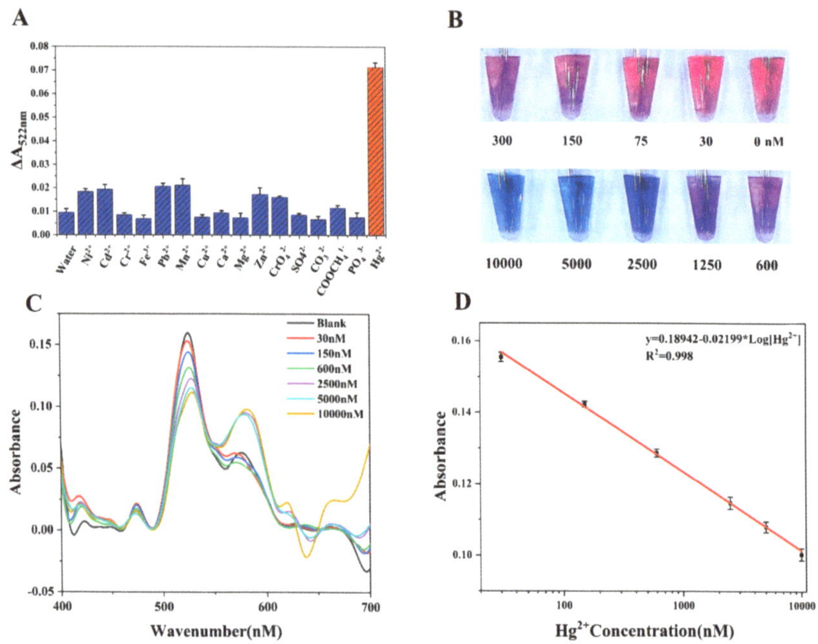

Figure 5. (**A**) Absorbance at 522 nm of AuNPs−GSH solutions when mixed with 16 different metal ions; (**B**) color changes caused by the addition of different concentrations of Hg^{2+} to the AuNPs−GSH solution; (**C**) absorption spectra of the solution in (**B**) when detected by the proposed sensor; (**D**) linear relationship between absorbance at 522 nm and the Hg^{2+} concentration.

Based on the work described above, it was determined that the colorimetric biosensor of the smartphone is capable of detecting Hg^{2+} in water by AuNPs-GSH. We then evaluated the quantitative detection effect of the sensor under optimized conditions. When the concentration of Hg^{2+} increased, the aggregation of the AuNPs in the solution also increased, which means that the absorbance of the solution near 522 nm decreased gradually as the Hg^{2+} concentration increased from 0 to 1×10^4 nM (Figure S8), and this change was accompanied by a red-shift phenomenon (Figure 5C). The LOD (3 sb/slope) of Hg^{2+} is 1.2 nM/0.24 ppb (Figure 5D), which is well under the international community's limit values of 5 nM (1 ppb) and 10 nM (2 ppb) for Hg^{2+} in drinking water. The change in the Hg^{2+} concentration and absorbance showed a good linear relationship within the detection range from 30 nM to 1×10^4 nM (Figure 5D), which indicates that the colorimetric biosensor system, based on the use of a smartphone as a miniature spectrometer as proposed in this paper, could be used effectively to perform quantitative Hg^{2+} detection. Furthermore, rapid detection and high sensitivity were both achieved within an extensive color variation range (red-purple-blue), even when the solutions were difficult to distinguish with the naked eye (Figure 5B). This verified the reliability of the sensor's colorimetric detection performance in practical applications.

3.3.4. Spiking Experiment in Actual Samples

When establishing the calibration model, the sample preparation solution was standard purified water prepared in the laboratory. However, the water-quality parameters for natural water bodies are not the same as for standard purified water. When the configuration solution is from a natural water body, the substances in the water may cause results to be inaccurate, so the detection and evaluation of actual water samples is essential. According to the three classifications of drinking water in China's national standard (GB19298-2014, GB5749-2006), we selected natural mineral water, pure water, and tap water as actual samples for the spiking experiment.

Analytical results in Table 1 showed that the recoveries varied from 97.43% to 102.98% in the spiked Hg^{2+} samples. The recovery values indicated that our proposed method could be used for highly accurate Hg^{2+} detection in actual samples from different drinking waters.

Table 1. Analytical results of samples.

Sample	Added (nM)	Found (nM)	Recovery (%)	RSD (%, n = 3)
Tap water	75	77.23	102.98	3.25
	300	299.07	99.68	1.43
	1250	1267.68	101.41	1.98
	5000	5123.52	102.65	2.02
Mineral water	75	75.575	100.76	3.58
	300	295.61	98.54	3.23
	1250	1217.87	97.43	2.61
	5000	4938.94	98.78	2.79
Pure water	75	74.91	100.14	1.06
	300	295.82	98.61	1.70
	1250	1277.33	102.18	1.86
	5000	4958.97	99.18	1.15

In addition, river water is an essential part of the ecological system and is an important source of drinking water, so we have also listed it as our test object. Due to the complex composition of river water, we left the samples standing overnight and diluted them five times with deionized water. For comparison, the inductively coupled plasma-mass spectrometry (ICP-MS) was also used to detect the spiked samples. As listed in Table S3, the results of the presented method were similar to those of ICP-MS, with the recoveries varying from 99.76% to 115.88%, which indicated that the proposed method showed excellent potential for detecting Hg^{2+} in complex, river water samples.

Furthermore, we compared the performance of the proposed strategy with those of other sensors for Hg^{2+} detection, based on AuNPs and on the use of a professional spectrometer or smartphone as the readout method, as shown in Table 2. First, we could obtain a detection limit with the same order of magnitude as those of precision instruments based on the Hg^{2+} detection of the smartphone. Second, when compared with the other methods presented, the Hg^{2+} detection method proposed in this paper showed a comparable linear range and faster response time. Third, combined with Table S1, we find that our method could work well with most cheap, mid- to low-end phones currently on the market. Additionally, compared to other smartphone-based methods, we can perform a more detailed analysis that incorporates spectral information rather than solely the data of color. Finally, through practical application to Hg^{2+} detection and by comparison with related work from other researchers, this paper demonstrated that the colorimetric biosensor with a smartphone as a low-cost micro-spectrometer has great potential for future applications, and that this sensor can perform colorimetric detection accurately, effectively, and stably within the visible light range.

Table 2. Comparison between the practical application effect of the proposed colorimetric biosensor to Hg^{2+} detection and the results of previously reported works.

Materials	Incubation Time(s)	Tool *	Linear Rangen(M)	Sample	LOD for Hg^{2+}	Ref.
AuNPs-MBT	300	S	50–10^3	Lake water	6.0 nM/1.20 ppb	[45]
AuNPs-CTAB	1800	S	20–10^3	Tap water	11.9 nM/2.39 ppb	[46]
AuNPs-MSA	300	S	10–10^4	Tap water	4.8 nM/0.96 ppb	[47]
AuNPs-AA	300	S	9–1.27×10^4	Tap water	8.8 nM/1.76 ppb	[48]
AuNPs-APTES	1200	S	15–92	River water	10 nM/2.01 ppb	[49]
AuNPs-DETL	900	S	100–5×10^3	River water	24.0 nM/4.81 ppb	[37]
AuNPs-DTT	480	P	54–267	Rain water	17.0 nM/3.40 ppb	[50]
AuNPs-H_2O_2	1800	P	100–10^4	Lake water	40.0 nM/8.02 ppb	[51]
AuNPs-GSH	5	P	30–10^3	River water	1.2 nM/0.24 ppb	**This work**

* Readout tool type. S: Spectrophotometer, P: Phone.

4. Conclusions

In this work, we built a colorimetric micro-spectrometer system that used a smartphone as a readout tool for the detection of Hg^{2+} in drinking water. The mobile terminal of the sensor processed the image received using a combination of the smartphone's camera and optical accessory, and ultimately achieved the desired spectral detection effect. In the evaluation of smartphone detection performance, effective detection of both the target light source and the liquid being tested were achieved, and the detection limit of AuNPs reached 0.14 μM. Additionally, when the samples were colorless or very similar in color, the smartphone detection method showed high sensitivity in capturing color changes. We also used the sensor to detect more complex sample-color changes. Therefore, we used AuNPs-GSH as a detection probe material to detect Hg^{2+} in water, and obtained a LOD of 1.2 nM. This method was also applied successfully in actual samples of natural mineral water, pure water, tap water, and river water, with a recovery rate range of 97.4~115.9%. Following a series of performance evaluations and practical applications, it was concluded that the sensor proposed in this paper could use the smartphone as a micro-spectrometer to collect the spectral signals of AuNPs and perform accurate, stable, and fast colorimetric detection with a highly sensitive readout capability. This sensor has tremendous potential for future application in low-cost biological colorimetric rapid detection.

Supplementary Materials: The following supporting information can be downloaded at: https://www.mdpi.com/article/10.3390/bios12111017/s1, Figure S1: The light source accessory (GoSpectro) is installed before the cell phone camera. The light beam A is first passed through a collimating lens B through a slit and is divided by a diffraction grating and a trigonal prism C to form a visible light band, which is received by the cell phone camera; Table S1: Smartphones related information; Figure S2: Detection of mercury lamp sources. (A) The influence of different exposure times on Huawei smartphone. (B) The results of the three phones can be very close after the calibration and adjustment of the exposure time; Table S2: Performance verification results after the smartphone was debugged, where each monochromatic diode was detected 20 times during the experiment at 2 s intervals; Figure S3: (A) Quantitative detection of AuNPs by the sensor in this paper; (B) Quantitative detection of AuNPs by commercial Avantes spectrometer; Figure S4: Size distribution and Zeta potential of AuNPs, AuNPs-GSH, and AuNPs-GSH+Hg^{2+}; Figure S5: Influence of various metal ions on the color of AuNPs-GSH solution; Figure S6: Influence of various metal ions on the absorbance of AuNPs-GSH solution; Figure S7: The change of absorbance in AuNPs-GSH solution after adding deionized water and metal ions mixture solution with/without Hg^{2+} (0.5 μM), respectively. A, AuNPs. M, metal ions mixture solution (Ni^{2+}, Cd^{2+}, Cr^{3+}, Pb^{2+}, Mn^{2+}, Cu^{2+}, $CrO_4{}^{2-}$, Ca^{2+}, Mg^{2+}, Zn^{2+}, and Fe^{3+} ions, 0.5 μM); Figure S8: The change in absorbance corresponds to the concentration of Hg^{2+}; Table S3: Determination of Hg^{2+} in river water by the proposed method.

Author Contributions: Y.G. and L.J.: conceptualization, methodology, validation, formal analysis, data curation, writing—original draft, visualization. F.C., X.L., Y.Z. and C.Y.: methodology, investigation. Z.G., M.Z. and P.L.: formal analysis, investigation. Y.H. and D.D.: conceptualization, data curation, writing—original draft, writing—review and editing, visualization, supervision, project administration, funding acquisition. All authors have read and agreed to the published version of the manuscript.

Funding: This research was financially supported by the Distinguished Young Scientists Program of Beijing Natural Science Foundation (JQ19023), Tsinghua Shenzhen International Graduate School Foundation (JC2022022), Science and Technology Innovation Ability Construction Program of Beijing Academy of agriculture and Forestry Sciences (KJCX20220405), and National Natural Science Foundation of China (31972148, 32101609, 32171627).

Institutional Review Board Statement: Not applicable.

Informed Consent Statement: Not applicable.

Data Availability Statement: Not applicable.

Conflicts of Interest: The authors declare that they have no known competing financial interests or personal relationships that could have appeared to influence the work reported in this paper.

References

1. Miretzky, P.; Cirelli, A.F. Hg(II) removal from water by chitosan and chitosan derivatives: A review. *J. Hazard. Mater.* **2009**, *167*, 10–23. [CrossRef] [PubMed]
2. Dave, N.; Chan, M.Y.; Huang, P.-J.J.; Smith, B.D.; Liu, J. Regenerable DNA-functionalized hydrogels for ultrasensitive, instrument-free mercury (II) detection and removal in water. *J. Am. Chem. Soc.* **2010**, *132*, 12668–12673. [CrossRef] [PubMed]
3. Mielcarek, K.; Nowakowski, P.; Puscion-Jakubik, A.; Gromkowska-Kepka, K.J.; Soroczynska, J.; Markiewicz-Zukowska, R.; Naliwajko, S.K.; Grabia, M.; Bielecka, J.; Zmudzinska, A.; et al. Arsenic, cadmium, lead and mercury content and health risk assessment of consuming freshwater fish with elements of chemometric analysis. *Food Chem.* **2022**, *379*, 132167. [CrossRef] [PubMed]
4. Llobet, J.M.; Falcó, G.; Casas, C.; Teixidó, A.; Domingo, J.L. Concentrations of Arsenic, Cadmium, Mercury, and Lead in Common Foods and Estimated Daily Intake by Children, Adolescents, Adults, and Seniors of Catalonia, Spain. *J. Agric. Food Chem.* **2003**, *51*, 838–842. [CrossRef] [PubMed]
5. Yang, Z.; Lu, W.; Long, Y.; Bao, X.; Yang, Q. Assessment of heavy metals contamination in urban topsoil from Changchun City, China. *J. Geochem. Explor.* **2011**, *108*, 27–38. [CrossRef]
6. Lima, E.A.; Cunha, F.A.S.; Junior, M.M.S.; Lyra, W.S.; Santos, J.C.C.; Ferreira, S.L.C.; Araujo, M.C.U.; Almeida, L.F. A fast and sensitive flow-batch method with hydride generating and atomic fluorescence spectrometric detection for automated inorganic antimony speciation in waters. *Talanta* **2020**, *207*, 119834. [CrossRef]
7. Goulle, J.P.; Mahieu, L.; Castermant, J.; Neveu, N.; Bonneau, L.; Laine, G.; Bouige, D.; Lacroix, C. Metal and metalloid multi-elementary ICP-MS validation in whole blood, plasma, urine and hair. Reference values. *Forensic Sci. Int.* **2005**, *153*, 39–44. [CrossRef]
8. Wang, Y.; Liu, G.; Li, Y.; Liu, Y.; Guo, Y.; Shi, J.; Hu, L.; Cai, Y.; Yin, Y.; Jiang, G. Occurrence of Mercurous [Hg(I)] Species in Environmental Solid Matrices as Probed by Mild 2-Mercaptoethanol Extraction and HPLC-ICP-MS Analysis. *Environ. Sci. Technol. Lett.* **2020**, *7*, 482–488. [CrossRef]
9. Kovac, M.; Bulaic, M.; Jakovljevic, J.; Nevistic, A.; Rot, T.; Kovac, T.; Dodlek Sarkanj, I.; Sarkanj, B. Mycotoxins, Pesticide Residues, and Heavy Metals Analysis of Croatian Cereals. *Microorganisms* **2021**, *9*, 216. [CrossRef]
10. Wu, S.; Yu, Q.; He, C.; Duan, N. Colorimetric aptasensor for the detection of mercury based on signal intensification by rolling circle amplification. *Spectrochim. Acta Part A Mol. Biomol. Spectrosc.* **2020**, *224*, 117387. [CrossRef]
11. Jeevika, A.; Shankaran, D.R. Functionalized silver nanoparticles probe for visual colorimetric sensing of mercury. *Mater. Res. Bull.* **2016**, *83*, 48–55. [CrossRef]
12. Zhang, L.; Xu, Y.; Xu, J.; Zhang, H.; Zhao, T.; Jia, L. Intelligent multicolor nano-sensor based on nontoxic dual fluoroprobe and MOFs for colorful consecutive detection of Hg^{2+} and cysteine. *J. Hazard. Mater.* **2022**, *430*, 128478. [CrossRef]
13. Ergun, E.G.C. Three in one sensor: A fluorometric, colorimetric and paper based probe for the selective detection of mercury(ii). *New J. Chem.* **2021**, *45*, 4202–4209. [CrossRef]
14. Wang, Q.; Wang, T.; Lin, H.; Meng, W.; Zhang, C.; Cai, P.; Hao, T.; Wu, Y.; Guo, Z. Disposable Faraday cage-type aptasensor for ultrasensitive determination of sub-picomolar Hg(II) via fast scan voltammetry. *Sens. Actuators B Chem.* **2020**, *320*, 128349. [CrossRef]
15. Palanisamy, S.; Thangavelu, K.; Chen, S.-M.; Velusamy, V.; Chang, M.-H.; Chen, T.-W.; Al-Hemaid, F.M.A.; Ali, M.A.; Ramaraj, S.K. Synthesis and characterization of polypyrrole decorated graphene/β-cyclodextrin composite for low level electrochemical detection of mercury (II) in water. *Sens. Actuators B Chem.* **2017**, *243*, 888–894. [CrossRef]
16. Tian, Y.; Hu, H.; Chen, P.; Dong, F.; Huang, H.; Xu, L.; Yan, L.; Song, Z.; Xu, T.; Chu, W. Dielectric Walls/Layers Modulated 3D Periodically Structured SERS Chips: Design. Batch Fabrication, and Applications. *Adv. Sci.* **2022**, *9*, e2200647. [CrossRef]

17. Liu, X.; Huang, D.; Lai, C.; Qin, L.; Zeng, G.; Xu, P.; Li, B.; Yi, H.; Zhang, M. Peroxidase-Like Activity of Smart Nanomaterials and Their Advanced Application in Colorimetric Glucose Biosensors. *Small* **2019**, *15*, e1900133. [CrossRef]
18. Liu, B.; Zhuang, J.; Wei, G. Recent advances in the design of colorimetric sensors for environmental monitoring. *Environ. Sci.* **2020**, *7*, 2195–2213. [CrossRef]
19. Moitra, P.; Alafeef, M.; Dighe, K.; Frieman, M.B.; Pan, D. Selective Naked-Eye Detection of SARS-CoV-2 Mediated by N Gene Targeted Antisense Oligonucleotide Capped Plasmonic Nanoparticles. *ACS Nano* **2020**, *14*, 7617–7627. [CrossRef]
20. Deng, J.; Yu, P.; Wang, Y.; Yang, L.; Mao, L.J.A.M. Visualization and quantification of neurochemicals with gold nanoparticles: Opportunities and challenges. *Adv. Mater.* **2014**, *26*, 6933–6943. [CrossRef]
21. Aldewachi, H.; Chalati, T.; Woodroofe, M.N.; Bricklebank, N.; Sharrack, B.; Gardiner, P. Gold nanoparticle-based colorimetric biosensors. *Nanoscale* **2017**, *10*, 18–33. [CrossRef] [PubMed]
22. Chang, C.C.; Chen, C.P.; Wu, T.H.; Yang, C.H.; Lin, C.W.; Chen, C.Y. Gold Nanoparticle-Based Colorimetric Strategies for Chemical and Biological Sensing Applications. *Nanomaterials* **2019**, *9*, 861. [CrossRef] [PubMed]
23. Song, M.; Khan, I.M.; Wang, Z. Research Progress of Optical Aptasensors Based on AuNPs in Food Safety. *Food Anal. Methods* **2021**, *14*, 2136–2151. [CrossRef]
24. Tan, L.; Zhang, Y.; Qiang, H.; Li, Y.; Sun, J.; Hu, L.; Chen, Z. A sensitive Hg(II) colorimetric sensor based on synergistic catalytic effect of gold nanoparticles and Hg. *Sens. Actuators B Chem.* **2016**, *229*, 686–691. [CrossRef]
25. Vidal, E.; Lorenzetti, A.S.; Garcia, C.D.; Domini, C.E. Use of universal 3D-Printed smartphone spectrophotometer to develop a time-based analysis for hypochlorite. *Anal. Chim. Acta* **2021**, *1151*, 338249. [CrossRef]
26. Lim, J.W.; Kim, T.Y.; Woo, M.A. Trends in sensor development toward next-generation point-of-care testing for mercury. *Biosens. Bioelectron.* **2021**, *183*, 113228. [CrossRef]
27. Poh, J.-J.; Wu, W.-L.; Goh, N.W.-J.; Tan, S.M.-X.; Gan, S.K.-E. Spectrophotometer on-the-go: The development of a 2-in-1 UV–Vis portable Arduino-based spectrophotometer. *Sens. Actuators A Phys.* **2021**, *325*, 56966. [CrossRef]
28. Jain, R.; Thakur, A.; Kaur, P.; Kim, K.-H.; Devi, P. Advances in imaging-assisted sensing techniques for heavy metals in water: Trends, challenges, and opportunities. *TrAC Trends Anal. Chem.* **2020**, *123*, 115758. [CrossRef]
29. Zheng, L.; Cai, G.; Wang, S.; Liao, M.; Li, Y.; Lin, J. A microfluidic colorimetric biosensor for rapid detection of Escherichia coli O157:H7 using gold nanoparticle aggregation and smart phone imaging. *Biosens. Bioelectron.* **2019**, *124–125*, 143–149. [CrossRef]
30. Bhalla, N.; Pan, Y.; Yang, Z.; Payam, A.F. Opportunities and Challenges for Biosensors and Nanoscale Analytical Tools for Pandemics: COVID-19. *ACS Nano* **2020**, *14*, 7783–7807. [CrossRef]
31. Fossum, E.R. CMOS image sensors: Electronic camera-on-a-chip. *IEEE Trans. Electron Devices* **1997**, *44*, 1689–1698. [CrossRef]
32. Dittrich, P.-G.; Bichra, M.; Stiehler, D.; Pfützenreuter, C.; Radtke, L.; Rosenberger, M.; Notni, G.; Messinger, D.W.; Velez-Reyes, M. Extended Characterization of Multispectral Resolving Filter-on-Chip Snapshot-Mosaic CMOS Cameras. In Proceedings of the Algorithms, Technologies, and Applications for Multispectral and Hyperspectral Imagery XXV, Baltimore, MA, USA, 16–18 April 2019.
33. Biswas, P.C.; Rani, S.; Hossain, M.A.; Islam, M.R.; Canning, J. Recent Developments in Smartphone Spectrometer Sample Analysis. *IEEE J. Sel. Top. Quantum Electron.* **2021**, *27*, 5500212. [CrossRef]
34. Ma, S.; Liu, Y.; Zheng, F.; Li, F.; Yu, D.; Xiao, A.; Yang, X. Development and Reliability study of 3D WLCSP for automotive CMOS image sensor using TSV technology. In Proceedings of the 2020 IEEE 70th Electronic Components and Technology Conference (ECTC), Orlando, FL, USA, 3–30 June 2020; pp. 461–466.
35. Frens, G. Controlled Nucleation for the Regulation of the Particle Size in Monodisperse Gold Suspensions. *Nat. Phys. Sci.* **1973**, *241*, 20–22. [CrossRef]
36. Tsogas, G.Z.; Kappi, F.A.; Vlessidis, A.G.; Giokas, D.L. Recent Advances in Nanomaterial Probes for Optical Biothiol Sensing: A Review. *Anal. Lett.* **2017**, *51*, 443–468. [CrossRef]
37. Huang, D.; Liu, X.; Lai, C.; Qin, L.; Zhang, C.; Yi, H.; Zeng, G.; Li, B.; Deng, R.; Liu, S.; et al. Colorimetric determination of mercury(II) using gold nanoparticles and double ligand exchange. *Mikrochim. Acta* **2018**, *186*, 31. [CrossRef]
38. Yoon, S.J.; Nam, Y.S.; Lee, Y.; Oh, I.H.; Lee, K.B. A dual colorimetric probe for rapid environmental monitoring of Hg^{2+} and As^{3+} using gold nanoparticles functionalized with d-penicillamine. *RSC Adv.* **2021**, *11*, 5456–5465. [CrossRef]
39. Duan, J.; Yin, H.; Wei, R.; Wang, W. Facile colorimetric detection of Hg^{2+} based on anti-aggregation of silver nanoparticles. *Biosens. Bioelectron.* **2014**, *57*, 139–142. [CrossRef]
40. Tolessa, T.; Tan, Z.Q.; Yin, Y.G.; Liu, J.F. Single-drop gold nanoparticles for headspace microextraction and colorimetric assay of mercury (II) in environmental waters. *Talanta* **2018**, *176*, 77–84. [CrossRef]
41. Najafzadeh, F.; Ghasemi, F.; Hormozi-Nezhad, M.R. Anti-aggregation of gold nanoparticles for metal ion discrimination: A promising strategy to design colorimetric sensor arrays. *Sens. Actuators B Chem.* **2018**, *270*, 545–551. [CrossRef]
42. Xu, H.; Wang, Y.; Huang, X.; Li, Y.; Zhang, H.; Zhong, X. Hg^{2+}-mediated aggregation of gold nanoparticles for colorimetric screening of biothiols. *Analyst* **2012**, *137*, 924–931. [CrossRef]
43. Wang, J.; Fang, X.; Cui, X.; Zhang, Y.; Zhao, H.; Li, X.; He, Y. A highly sensitive colorimetric probe for Cd^{2+}, Hg^{2+} and ascorbic acid determination based on trithiocyanuric acid-AuNPs. *Talanta* **2018**, *188*, 266–272. [CrossRef] [PubMed]
44. Liu, D.; Qu, W.; Chen, W.; Zhang, W.; Wang, Z.; Jiang, X. Highly sensitive, colorimetric detection of mercury(II) in aqueous media by quaternary ammonium group-capped gold nanoparticles at room temperature. *Anal. Chem.* **2010**, *82*, 9606–9610. [CrossRef] [PubMed]

45. Jin, W.; Huang, P.; Wei, G.; Cao, Y.; Wu, F. Visualization and quantification of Hg^{2+} based on anti-aggregation of label-free gold nanoparticles in the presence of 2-mercaptobenzothiazole. *Sens. Actuators B Chem.* **2016**, *233*, 223–229. [CrossRef]
46. Sun, X.; Liu, R.; Liu, Q.; Fei, Q.; Feng, G.; Shan, H.; Huan, Y. Colorimetric sensing of mercury (II) ion based on anti-aggregation of gold nanoparticles in the presence of hexadecyl trimethyl ammonium bromide. *Sens. Actuators B Chem.* **2018**, *260*, 998–1003. [CrossRef]
47. Xiong, X.; Lai, X.; Liu, J. Mercaptosuccinic acid-coated NIR-emitting gold nanoparticles for the sensitive and selective detection of Hg^{2+}. *Spectroch. Acta A Mol. Biomol. Spectrosc.* **2018**, *188*, 483–487. [CrossRef]
48. Liu, Y.; Liu, Y.; Xu, L.; Li, J.; Liu, X.; Liu, J.; Li, G. Highly selective, colorimetric detection of Hg^{2+} based on three color changes of AuNPs solution from red through sandy beige to celandine green. *Sens. Actuators B Chem.* **2017**, *249*, 331–338. [CrossRef]
49. Xie, Y. Colorimetric determination of Hg(II) via the gold amalgam induced deaggregation of gold nanoparticles. *Mikrochim. Acta* **2018**, *185*, 351. [CrossRef]
50. Motalebizadeh, A.; Bagheri, H.; Asiaei, S.; Fekrat, N.; Afkhami, A. New portable smartphone-based PDMS microfluidic kit for the simultaneous colorimetric detection of arsenic and mercury. *RSC Adv.* **2018**, *8*, 27091–27100. [CrossRef]
51. Li, X.; Wang, H.; Wang, X.; Lai, G.; Gong, J. "All-in-One" ultra-portable colorimetric sensor coupled with confinement effect for Hg^{2+} highly sensitive and instant detection. *Sens. Actuators B Chem.* **2022**, *351*, 130945. [CrossRef]

biosensors

MDPI

Article

RhB@MOF-5 Composite Film as a Fluorescence Sensor for Detection of Chilled Pork Freshness

Jingyi Li [1,2,3,4], Ning Zhang [5], Xin Yang [1,2,3,4], Xinting Yang [2,3,4], Zengli Wang [1,*] and Huan Liu [2,3,4,*]

1 College of Food Science and Nutritional Engineering, China Agricultural University, Beijing 100083, China; lijingyi@cau.edu.cn (J.L.); yangxincyn@163.com (X.Y.)
2 Research Center of Information Technology, Beijing Academy of Agricultural and Forestry Sciences, Beijing 100097, China; xingtingyang@nercita.org.cn
3 National Engineering Research Center for Information Technology in Agriculture, Beijing 100097, China
4 National Engineering Laboratory for Agri-Product Quality Traceability, Beijing 100097, China
5 Beijing Key Laboratory of Flavor Chemistry, Beijing Technology and Business University (BTBU), Beijing 100048, China; zh_ningts@btbu.edu.cn
* Correspondence: wangzengli@cau.edu.cn (Z.W.); liuhuan@nercita.org.cn (H.L.); Tel.: +86-10-62737066 (Z.W.); +86-10-51503630 (H.L.)

Abstract: This study presents a novel composite thin film based on rhodamine B encapsulated into MOF-5 (Metal Organic Frameworks) as a fluorescence sensor for the real-time detection of the freshness of chilled pork. The composite film can adsorb and respond to the volatile amines produced by the quality deterioration of pork during storage at 4 °C, with the fluorescence intensity of RhB decreasing over time. The quantitative model used for predicting the freshness indicator (total volatile base nitrogen) of pork was built using the fluorescence spectra (excited at 340 nm) of the RhB@MOF-5 composite film combined with the partial least squares (PLS) algorithm, providing R_c^2 and R_p^2 values of 0.908 and 0.821 and RMSEC (root mean square error of calibration) and RMSEP (root mean square error of prediction) values of 3.435 mg/100 g and 3.647 mg/100 g, respectively. The qualitative model established by the partial least squares discriminant analysis (PLS-DA) algorithm was able to accurately classify pork samples as fresh, acceptable or spoiled, and the accuracy was 86.67%.

Keywords: metal–organic frameworks; rhodamine B; volatile compounds; fluorescence sensing; chilled pork; freshness

Citation: Li, J.; Zhang, N.; Yang, X.; Yang, X.; Wang, Z.; Liu, H. RhB@MOF-5 Composite Film as a Fluorescence Sensor for Detection of Chilled Pork Freshness. *Biosensors* **2022**, *12*, 544. https://doi.org/10.3390/bios12070544

Received: 23 June 2022
Accepted: 18 July 2022
Published: 20 July 2022

1. Introduction

Pork is a highly nutritious food rich in protein and several essential amino acids and vitamins ideal for human consumption. However, fresh pork is a perishable food and can easily deteriorate under the action of enzymes and microorganisms in cold-chain logistics, which affects food safety [1]. Traditionally, the freshness of chilled pork has been evaluated based on chemical and microbial indicators, such as the total volatile base nitrogen (TVB-N) value and total viable count (TVC) [2]. However, these traditional detection methods have the disadvantages of being time-consuming and requiring cumbersome pretreatments, high operational requirements, and sample destruction. When meat deteriorates, volatile compounds are produced due to protein degradation and lipid oxidation. Therefore, if these volatile gases can be identified, pork freshness can be monitored in real time.

Electronic nose technology, which is often used to analyze food odors and can compensate for the deficiencies caused by the human subjectivity inherent in sensory evaluation, has been used to determine meat freshness [3,4]. However, as metal oxide sensors in the electronic nose often require the stimulation of electron transfer at a high temperature to produce a gas-sensitive response, these can easily be affected by environmental factors, limiting their application [5]. Cold-chain logistics are now developing from using traditional static methods for detecting food quality to using the "Internet of Things" to provide comprehensive, real-time monitoring [6]. Electronic nose devices are large and expensive, making the rapid sharing of information difficult. More recently, gas sensors based on pigments, which can perform detection and analysis based on the color change of pigments (such as anthocyanin and bromophenol blue) caused by changes in the pH of volatile substances arising in meat, have been developed [7–9]. However, these gas sensors have some disadvantages [10], such as pigment instability, a short shelf-life, and low sensitivity. Therefore, developing a rapid and reliable method based on new synthetic gas-sensing materials for identifying the volatile compounds of pork during chilled storage in real time, thus providing a freshness detection technology, has great significance and wide prospects for application in food preservation [11].

Metal–organic frameworks (MOFs) are a type of inorganic, porous, and coordinated polymer material formed from a central metal component and rigid ligands. These highly ordered MOFs are characterized by superior performance resulting from their diversified structure, ultra-high porosity, good adsorption, and surface modifiability, which can play a role in preconcentrating gas molecules and providing sufficient contact with analytes [12–14]. With their unique cage and channel-type porous structures, MOFs can efficiently assemble molecules with identifiable characteristics. Furthermore, owing to their greater specific surface area, these can provide a greater number of active sites, leading to more sensitive chemical sensing [15,16]. One type of MOF, $Zn_4(\mu_4\text{-}O)\text{-}(\mu_4\text{-}4\text{-carboxyl-3,5-dimethyl-4-carboxyl-pyrazole})_3$, containing carboxyl groups and coordinated unsaturated metal sites, can selectively capture harmful volatile organic compounds (VOCs) such as sarin and mustard gas [17]. Koh et al. designed a plasmonic nose based on an MOF-encapsulated Ag nanocube array that was able to identify and quantify several harmful VOCs and polycyclic aromatic hydrocarbons (PAHs) at the ppm level [18].

Introducing guest molecules or units, such as dye molecules or chromophores, into MOFs can improve their fluorescence properties, allowing them to be used as a luminous platform for the detection of target substances [19,20]. Yassine et al. used thin films of a rare earth metal (RE)-based MOF with an fcu topology as a fluorescence sensor to detect H_2S at room temperature [21]. The innovative material Rho@CZJ-3 (a luminescent MOF material containing 1D nanotube channels) was synthesized to probe different volatile organic molecules (VOMs) by modulating the energy transfer efficacy between two different emissions [22]. Combining the highly sensitive and selective identification offered by fluorescence spectroscopy with the gas adsorption properties of MOFs shows potential for producing new gas-sensing materials for practical applications.

This study aimed to synthesize a luminescent rhodamine B (RhB)@MOF-5 gas-sensing thin film for the real-time detection of the freshness of chilled pork (Scheme 1). First, RhB@MOF-5 was prepared from a mixture of zinc nitrate hexahydrate and terephthalic acid in a 2:1 ratio by adding RhB at a concentration of 2×10^{-3} mol/L, and the composite film was manufactured by mixing RhB@MOF-5 with polyvinylidene fluoride (PVDF). Next, changes in the volatile components during pork deterioration were analyzed using electronic nose technology, and the response of RhB@MOF-5 to these volatile compounds was discussed. Finally, the relationship between the fluorescence properties of the RhB@MOF-5 composite films and the TVB-N value of pork during storage at 4 °C was determined, and these data, in combination with chemometrics algorithms, were used to establish quantitative and qualitative models for the evaluation of pork freshness.

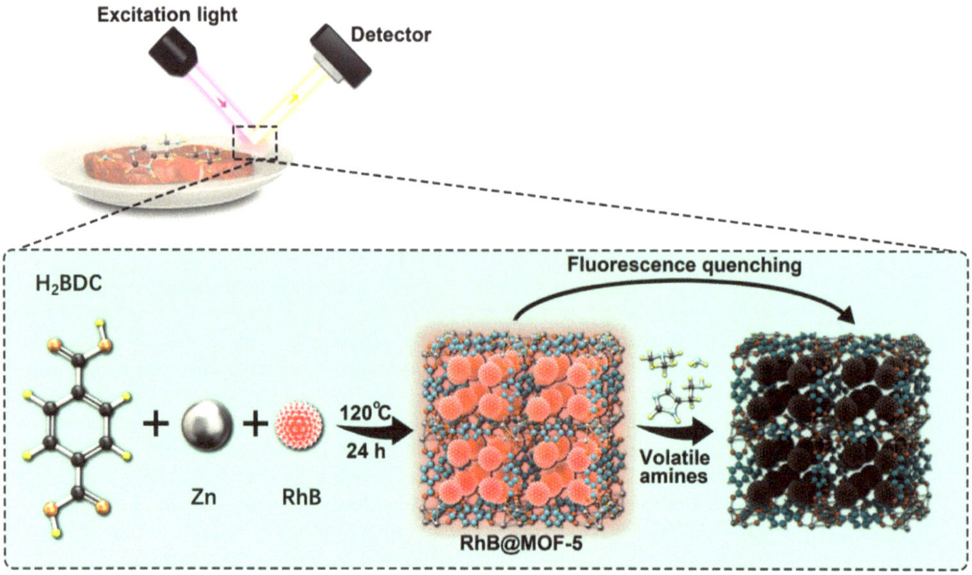

Scheme 1. Schematic illustration of the RhB@MOF-5 composite film for pork freshness detection.

2. Materials and Methods

2.1. Materials

Pork samples of *longissimus dorsi* muscles were obtained from the Beijing Ershang Dahongmen Meat Food Co., Ltd. (Beijing, China). The muscles were divided into approximately 50 g pieces on a sterile surface and placed on a tray. The RhB@MOF-5 composite films were made into uniform size labels, numbered, and then attached to plastic wrap on the tray. Thirty prepared pork samples were stored in an incubator at 4 °C.

Zinc nitrate hexahydrate ($Zn(NO_3)_2{\cdot}6H_2O$), 1,4-benzenedicarboxylic acid (H_2BDC), *N,N*-dimethylformamide (DMF), and rhodamine B (RhB) were purchased from Aladdin Reagent Company (Shanghai, China).

2.2. Preparation of RhB@MOF-5 Composite Film

RhB (38.3 mg) was added to dimethyl formamide (DMF, 40 mL) containing $Zn(NO_3)_2{\cdot}6H_2O$ (2.00 mmol) and H_2BDC (1.00 mmol),and treated with ultrasound for 60 min. The reaction mixture was then transferred to a Teflon-lined reactor and heated in an oven at 120 °C for 24 h. The mixture was then cooled to room temperature and separated by centrifugation (2500 rpm, 5 min), and the products were washed repeatedly with DMF until the supernatant exhibited no fluorescence. After further centrifugation, the products were dried under vacuum at 120 °C for 12 h to obtain the MOF powder [23]. The MOF and PVDF powders were mixed in a 4:5 ratio and then DMF was added. After dispersion using ultrasound for 30 min, the mixture was stirred overnight with a magnetic stirrer to obtain a light pink uniform viscous liquid. This liquid was poured onto a tray in a thin layer and then placed in an oven at 140 °C for about 30 min. After cooling naturally, the films were cut into 1 cm square pieces with a thickness of 0.1 mm and then placed in a dryer

2.3. Instrumentation

Scanning electron microscopy (SEM) images were obtained using the FEI Inspect F50 scanning electron microscope. Powder X-ray diffraction (PXRD) patterns were created using the Panalytical X-ray Diffractometer Smartlab-9 kW. Fourier transform infrared (FTIR) spectra were investigated using a Thermo Scientific Nicolet IS5 FTIR spectrophotometer within a range of 4000 to 400 cm^{-1}. The fluorescence spectra were obtained using a HITACHI F-7000 fluorescence spectrophotometer with a solid sample holder enabling the non-destructive analysis. Electronic nose detection was performed using the AlphaMOS Fox 4000 electronic nose with the test parameters as follows: balance temperature of the sample, 37 °C; balance time, 600 s; temperature of the sampler needle, 47 °C; clean and dry air used as the carrier gas at a flow rate of 150 mL/min; sampler volume, 1500 μL; delay time, 10 min; and five samples being processed at one time. The maximum response value of the sensor was taken as the characteristic value.

2.4. *Storage Stablity Experiment*

The RhB@MOF-5 composite films prepared in Section 2.2 were placed in a dark environment at 4 °C. Then, the fluorescence data were collected on day 0, day 30, day 45 and day 60. The fluorescence intensity on day 0 was I_0, and the ratio between day *n* and day 0 was I_n/I_0.

2.5. *Standard Measurement of Freshness Indicator*

According to the Chinese national standard (GB 5009.228-2016), the TVB-N value can be used as an indicator of meat freshness based on the collection of alkaline volatile nitrogen-containing substances, such as ammonia and amine [24]. Using the Kjeldahl method, the TVB-N value of chilled pork samples was determined on days 1, 2, 3, 4, 6, 8, 11, 13, 15, and 17 of storage, with three pork samples used for each determination.

2.6. *Statistical Analysis*

The data were analyzed using IBM SPSS Statistics (Version 26, IBM Corp., Armonk, NY, USA) and graphs were plotted using Origin 2018 (OriginLab Corp., Northampton, MA, USA).

Partial least squares (PLS) regression was used to identify quantitative correlations between the spectral data and the freshness indicator, using Unscrambler X 10.0.1 (CAMO PROCESS AS, Oslo, Norway). The performance of the models was assessed using the square of the correlation coefficient (R^2), the root mean square error of calibration (RMSEC), and the root mean square error of prediction (RMSEP). The models will generally give higher ${R_c}^2$ and ${R_p}^2$ values and lower RMSEC and RMSEP values. When R^2 is closer to 1 and the R_c and R_p values are more similar, the predictions made by the models are usually more accurate [25,26].

Partial least squares discriminant analysis (PLS-DA) is a linear classification method. It can combine the characteristics of PLS regression with the discrimination ability of a classification technique and is used to find mathematical models that can identify which class each sample belongs to [27]. PLS-DA models were applied using SIMCA-P 11.5 (Umetrics, Umeå, Sweden), which allowed the graphical visualization and understanding of the different data patterns and relationships using the scores and loadings of latent variables (LVs).

3. Results and Discussion

3.1. *Synthesis and Characterization of RhB@MOF-5*

3.1.1. Selection of RhB Concentration

Images of MOF-5, RhB, RhB mixed with MOF-5, and RhB@MOF-5 with different RhB concentrations under visual light and 365 nm UV light are shown in Figure 1a,b. The luminescence intensity of RhB@MOF-5 was much higher than that of pure RhB and MOF-5 powder. Fluorescent dyes often undergo fluorescence quenching due to aggregation, but the porous structure of MOFs can act as a "solid solvent" [19]. According to the related reference [28], the pore size of MOF-5 is around 1.56 nm. The immobilization of dyes into the pore spaces of MOFs can disperse the dye molecules and minimize aggregation-induced quenching, resulting in increased luminescence. Furthermore, the photoluminescence properties of RhB@MOF-5 were clearly different from those of the mixture of MOF-5 and RhB, which indicated that RhB had been enclosed in the pores of MOF-5.

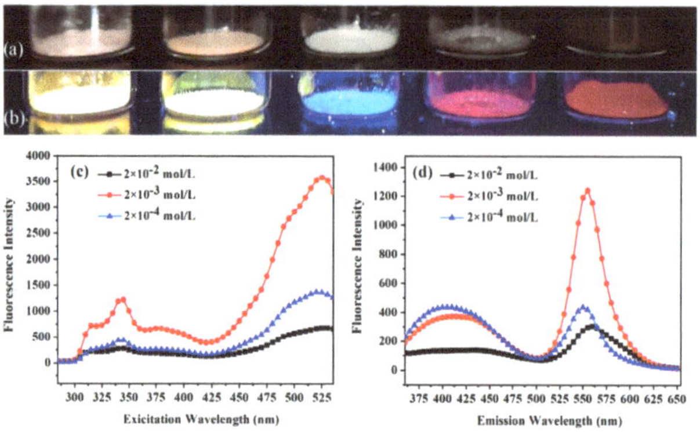

Figure 1. (**a**) Images of RhB@MOF-5 (RhB at a concentration of 2×10^{-3} mol/L), RhB@MOF-5 (RhB at a concentration of 2×10^{-2} mol/L), MOF-5, RhB mixed with MOF-5 and RhB under visual light. (**b**) Images of RhB@MOF-5 (RhB at a concentration of 2×10^{-3} mol/L), RhB@MOF-5 (RhB at a concentration of 2×10^{-2} mol/L), MOF-5, RhB mixed with MOF-5 and RhB under 365 nm UV light. (**c**) Fluorescence spectra (emission at 550 nm) for RhB@MOF-5 (RhB at concentrations of 2×10^{-2}, 2×10^{-3}, and 2×10^{-4} mol/L). (**d**) Fluorescence spectra (excitation at 340 nm) for RhB@MOF-5 (RhB at concentrations of 2×10^{-2}, 2×10^{-3}, and 2×10^{-4} mol/L).

Previous research has shown that the fluorescence emission wavelength of RhB is about 550 nm [29]. To determine the optimal RhB concentration, fluorescence spectra of RhB@MOF-5 powder samples with different RhB concentrations (including 2×10^{-2}, 2×10^{-3}, and 2×10^{-4} mol/L) were measured with emission at 550 nm, as shown in Figure 1c. The fluorescence intensities exhibited maximum values when the RhB concentration was 2×10^{-3} mol/L. When the dye concentration is low, the fluorescence intensity increases with increasing dye concentration. However, when the dye concentration reaches a certain value, its aggregation will be enhanced, which increases the nonradiative energy transfer between molecules but decreases the fluorescence intensity [30]. MOF-5 provides abundant pores that allow RhB to be well dispersed, which decreases aggregation and thus improves the fluorescence properties. Figure 1d also shows the redshift in the emission wavelength as the concentration of RhB increased. This might be due to energy transferring between the component molecules when RhB aggregated into a dimer or polymer form. Therefore, the RhB/DMF solution with a concentration of 2×10^{-3} mol/L was selected to prepare RhB@MOF-5 for the following experiments.

3.1.2. Characterization of RhB@MOF-5

The morphological and structural characterization of RhB@MOF-5 was conducted using the following tests. First, SEM images of MOF-5 and RhB@MOF-5 were analyzed, as shown in Figure 2a,b. The size of MOF-5 was 60–80 μm, while the size of RhB@MOF-5 was about 80 μm. The surface of RhB@MOF-5 was slightly rougher than that of MOF-5. RhB@MOF-5 is a square crystal with a similar structure to that of MOF-5, indicating that RhB encapsulation did not change the morphology of MOF-5. The powder X-ray diffraction (PXRD) characterization of RhB@MOF-5 was also conducted, as shown in Figure 2c. This showed that RhB encapsulation did not obviously change the diffraction peak position or reduce the intensity of MOF-5. Therefore, RhB@MOF-5 retained the structure of MOF-5. Moreover, the XRD peaks of MOF-5 and RhB@MOF-5 were similar, indicating that RhB molecules had been enclosed in the pores rather than being physically adsorbed on the surface of MOF-5, and that the composite material had been successfully prepared. Figure 2d shows the FT-IR spectra of MOF-5 and RhB@MOF-5, which were highly similar, with no shift in the symmetric and asymmetric stretching vibrations of carboxylic acid at 1598 and 1390 cm^{-1}, respectively, indicating that RhB encapsulation did not affect the coordination between H_2BDC and zinc(II) ions.

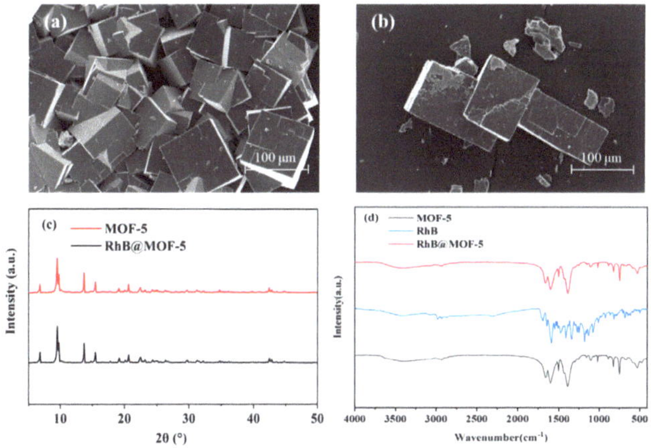

Figure 2. (**a**) SEM image of MOF-5; (**b**) SEM image of RhB@MOF-5; (**c**) PXRD patterns of MOF-5 and RhB@MOF-5; and (**d**) FT-IR spectra of MOF-5, RhB and RhB@MOF-5.

3.2. Response of RhB@MOF-5 to Volatile Compounds during Pork Deterioration

3.2.1. Analysis of Characteristic Volatile Components by Electronic Nose

The electronic nose is an odor analysis system based on a gas sensor array, with the different sensors able to identify the various volatile components. Figure 3 shows the changes in the responses of the 18 sensors of the electronic nose to volatile compounds from pork stored at 4 °C (Table S1). As the storage time increased, the response values of the electronic nose sensors also changed, with sensors LY2/G, LY2/AA, LY2/GH, and LY2/gCTL showing particularly obvious changes. Sensor LY2/G responded to changes in ammonia, amine compounds, ketones and alcohols; sensor LY2/AA responded to changes in ethanol, acetone, and ammonia; sensor LY2/GH responded to changes in ammonia and amine compounds; and sensor LY2/gCTL responded to changes in H_2S. Overall, this showed that the contents of compounds including ammonia, amines, ketones, and alcohols changed a great deal as pork deteriorated during chilled storage.

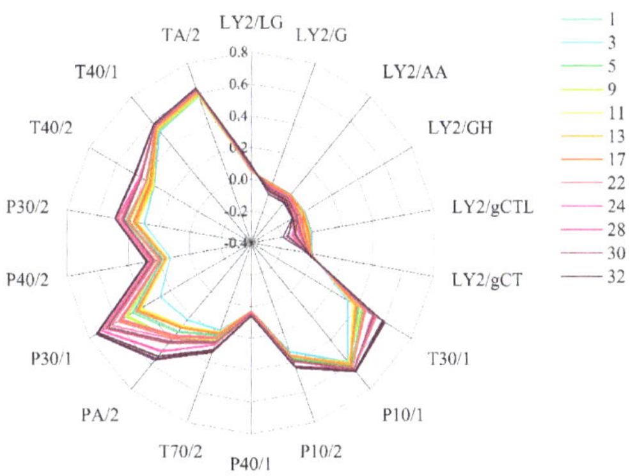

Figure 3. Changes in the response of 18 sensors of the electronic nose to pork samples stored at 4 °C (day 1, 3, 5, 9, 11, 13, 17, 22, 24, 28, 30 and 32).

3.2.2. Infrared Analysis of RhB@MOF-5 Response to Pork Deterioration

Infrared analysis can be used to characterize the chemical bonds and functional groups of substances to provide specific fingerprinting characteristics. FT-IR spectra of RhB@MOF-5 were collected after the adsorption of volatile components produced by pork deterioration, as shown in Figure 4. Characteristic bands of the symmetric and asymmetric stretching vibrations of terephthalic acid were still present at 1598 and 1390 cm^{-1}, respectively. The intensity of the Zn–O band at 530 cm^{-1}, the C-H band at 750 cm^{-1} and the C=O band at 1015 cm^{-1} decreased, showing that the structure of RhB@MOF-5 had been affected. A distinct, wide band appeared from 2800 to 3500 cm^{-1} and a sharp band appeared at 3600 cm^{-1}, which were both attributed to O-H stretching, showing that RhB@MOF-5 had absorbed water. Furthermore, a N-H stretching vibration band was observed at 3243 cm^{-1} and a C-N stretching vibration band appeared at 1310 cm^{-1}, which indicated that RhB@MOF-5 adsorbed volatile amines produced by the quality deterioration of pork during storage.

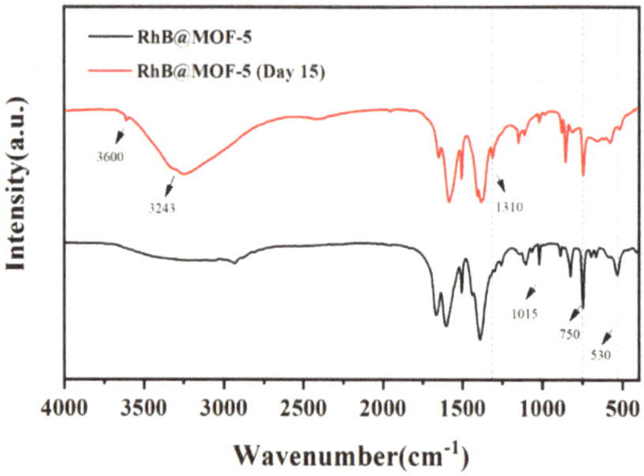

Figure 4. FT-IR spectra of RhB@MOF-5 response to pork deterioration.

3.2.3. Fluorescence Sensing Analysis of RhB@MOF-5 Response to Pork Deterioration

Three-dimensional fluorescence spectra can provide comprehensive information on components in a complex system. Four fluorescence peaks were identified in the 3-D florescence spectra of RhB@MOF-5, as shown in Figure 5, Ex/Em peaks located at 330 nm/435 nm and 365 nm/440 nm were attributed to MOF-5, while Ex/Em peaks of 340 nm/550 nm and 520 nm/550 nm were characteristic peaks of RhB. As shown in Figure 5d, when RhB@MOF-5 adsorbed volatile compounds from pork deterioration, the fluorescence intensities of the two characteristic peaks of RhB (340 nm/550 nm and 520 nm/550 nm) greatly decreased. The RhB molecules were distributed in the pores of MOF-5 mainly through electrostatic action. When volatile amines diffused and adsorbed into the pores of MOF-5 and made contact with RhB, the electron cloud density of RhB molecules changed, resulting in a decrease in fluorescence intensity. These results demonstrated the feasibility of using RhB@MOF-5 as a gas-sensing material for monitoring the quality deterioration and detecting the freshness of chilled pork.

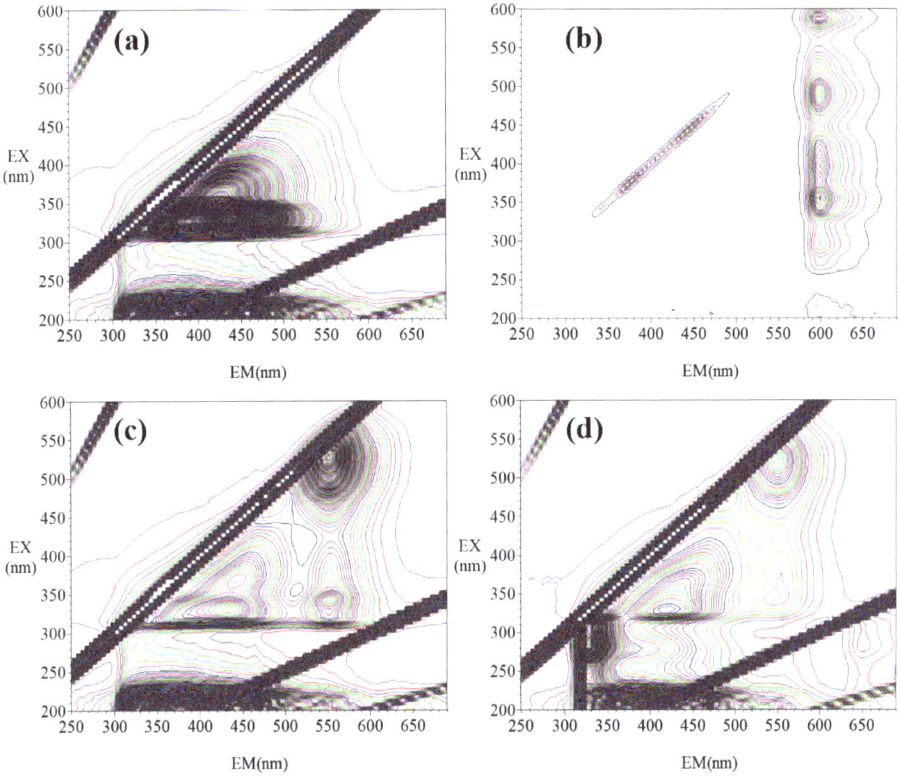

Figure 5. (**a**) Three-dimensional fluorescence spectrum of MOF-5; (**b**) 3-D fluorescence spectrum of RhB; (**c**) 3-D fluorescence spectrum of RhB@MOF-5; and (**d**) 3-D fluorescence spectrum of RhB@MOF-5 response to pork deterioration.

3.3. Application of RhB@MOF-5 Composite Film to Detecting Chilled Pork Freshness

To broaden the applications of MOFs and weaken the effect of their inherent vulnerability, a composite film was prepared using RhB@MOF-5 and PVDF powders at a 4:5 ratio (*w/w*) using the mixed matrix method, which simultaneously provided the mechanical flexibility of a polymer matrix and the high porosity of MOFs. A storage test showed that the composite film maintained its fluorescence stability for more than 60 d in a dark environment at 4 °C, as shown in Figure 6c.

Fluorescence wavelength scanning using a fixed excitation or emission wavelength has the advantage of simplifying spectral analysis and reducing the influence of light scattering, and only takes a few seconds per sample (3-D fluorescence spectroscopy requires more than 5 min for each scan). As shown in Figure 5c, scanning the RhB@MOF-5 composite film at an excitation wavelength of 340 nm facilitated simultaneous observation of the changes in characteristic peaks of both the MOF-5 (Ex/Em at 330 nm/435 nm) and RhB (Ex/Em at 340 nm/550 nm), allowing the properties of RhB@MOF-5 to be comprehensively determined. Figure 6a shows the fluorescence emission spectra of RhB@MOF-5 composite films (excited at 340 nm) during pork storage at 4 °C. Two characteristic peaks were identified at emission wavelengths near 420 nm and 550 nm, which were associated with MOF-5 and RhB, respectively. The intensities of these two peaks decreased significantly with increasing storage time. Furthermore, Figure 6b shows the fluorescence images of composite RhB@MOF-5 films (under excitation light of 530 nm) before and after pork spoilage, showing that the fluorescence intensity had clearly decreased. Overall, the volatile compounds produced by pork deterioration greatly affected the fluorescence properties of the composite film.

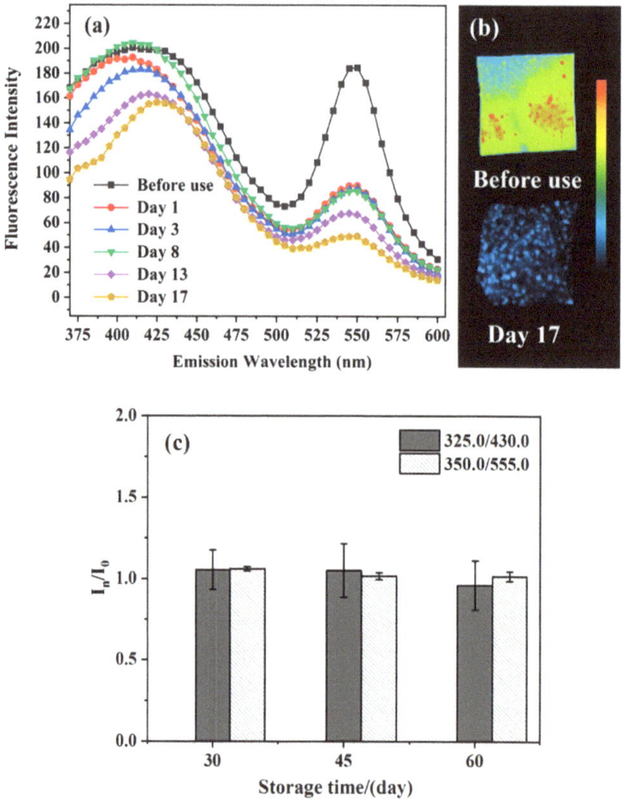

Figure 6. (**a**) Fluorescence emission spectra (excitation at 340 nm) of RhB@MOF-5 composite films of pork samples stored at 4 °C. (**b**) Fluorescence image of RhB@MOF-5 composite film before and after pork spoilage (excitation light of 530 nm). (**c**) Storage stability of RhB@MOF-5 composite film at 4 °C in dark conditions.

The TVB-N value can be used as an indicator of meat freshness based on the collection of alkaline volatile nitrogen-containing substances, such as ammonia and amine. The TVB-N values of the pork samples were measured on days 1, 2, 3, 4, 6, 8, 11, 13, 15, and 17 of chilled storage at 4 °C, and ranged between 11.17 and 54.68 mg/100 g, with an average value of 19.18 mg/100 g. Partial least squares (PLS) regression was used to establish a quantitative model to identify the relationship between the fluorescence spectral data of RhB@MOF-5 composite films and the TVB-N values of pork samples. The data matrix from 30 pork samples was randomly divided into a calibration set and a prediction set at a ratio of ~3:1 (23 in the correction set and 7 in the prediction set). Figure 7a shows the quantitative modeling results, with R_c^2 and R_p^2 values of 0.908 and 0.821, respectively, and RMSEC (root mean square error of calibration) and RMSEP (root mean square error of prediction) values of 3.435 and 3.647 mg/100 g, respectively. With both the R_c^2 and R_p^2 values being greater than 0.8, and the RMSEC and RMSEP values being similar, the PLS regression model was considered to be effective with a good prediction accuracy.

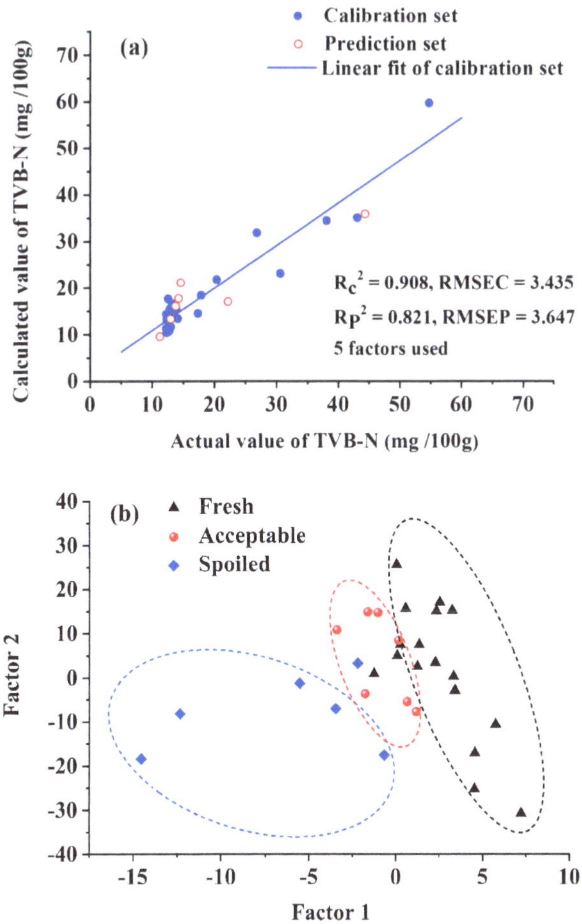

Figure 7. (**a**) The results of PLS modeling of the TVB-N values determined for pork samples during chilled storage. (**b**) The latent variable scores of PLS-DA modeling of pork samples (classified as fresh, acceptable or spoiled).

According to Chinese national food safety standards for meat quality, the TVB-N values of fresh pork is below 15.0 mg/100 g, of acceptable pork is around 15.0–25.0 mg/100 g, and of spoiled pork is above 25.0 mg/100 g [24], which serves as a reference for establishing a qualitative model based on fluorescence spectra of RhB@MOF-5 film combined with the PLS-DA algorithm. Figure 7b shows the visualization of the latent variable score plot map of the qualitative model, as defined by the principal components 1 and 2, and the three classes of pork freshness (fresh, acceptable and spoiled) can be clearly distinguished with an accuracy of 86.67%, which can be used to estimate the freshness classifications of pork samples reliably.

Overall, the RhB@MOF-5 composite film has the potential to be used as a fluorescence gas sensor for the sensitive, rapid, and nondestructive detection and visual monitoring of pork quality deterioration and freshness during refrigerated storage.

4. Conclusions

This study developed a RhB@MOF-5 composite thin film that can be used as an intelligent label in pork packaging, enabling the sensitive and nondestructive detection of pork freshness during chilled storage based on fluorescence sensing. Changes in the FT-IR spectra of RhB@MOF-5 after exposure to deteriorating pork showed that this material adsorbed volatile amines. The fluorescence intensity of the RhB@MOF-5 composite film tended to weaken with increasing storage time. Quantitative models for predicting the freshness indicator and qualitative models for the classification of pork freshness both produced satisfactory results. This study provides the first examination of the dye@MOF system as a gas-sensing material for meat freshness evaluation, showing its potential for application in food quality detection.

Supplementary Materials: The following supporting information can be downloaded at: https://www.mdpi.com/article/10.3390/bios12070544/s1, Table S1: Eighteen channels in electronic nose.

Author Contributions: Conceptualization: H.L. and J.L.; methodology: J.L.; software: N.Z. and X.Y. (Xin Yang); validation: X.Y. (Xinting Yang); formal analysis: N.Z. and Z.W.; investigation: H.L.; resources: Z.W., H.L. and X.Y. (Xinting Yang). All authors have read and agreed to the published version of the manuscript.

Funding: This study was financially supported by the National Natural Science Foundation of China (32001775), the Open Project Program of Beijing Key Laboratory of Flavor Chemistry, and the Beijing Young Scholars Program.

Institutional Review Board Statement: Not applicable.

Informed Consent Statement: Not applicable.

Data Availability Statement: The data presented in this study are available upon request from the corresponding author.

Conflicts of Interest: The authors declare no conflict of interest.

References

1. Bruckner, S.; Albrecht, A.; Petersen, B.; Kreyenschmidt, J. Influence of Cold Chain Interruptions on the Shelf Life of Fresh Pork and Poultry: Influence of Cold Chain Interruptions. *Int. J. Food Sci. Technol.* **2012**, *47*, 1639–1646. [CrossRef]
2. Li, N.; Zhang, Y.; Wu, Q.; Gu, Q.; Chen, M.; Zhang, Y.; Sun, X.; Zhang, J. High-Throughput Sequencing Analysis of Bacterial Community Composition and Quality Characteristics in Refrigerated Pork during Storage. *Food Microbiol.* **2019**, *83*, 86–94. [CrossRef] [PubMed]
3. Tian, X.-Y.; Cai, Q.; Zhang, Y.-M. Rapid Classification of Hairtail Fish and Pork Freshness Using an Electronic Nose Based on the PCA Method. *Sensors* **2011**, *12*, 260–277. [CrossRef] [PubMed]
4. Li, M.; Wang, H.; Sun, L.; Zhao, G.; Huang, X. Application of Electronic Nose for Measuring Total Volatile Basic Nitrogen and Total Viable Counts in Packaged Pork During Refrigerated Storage: E-Nose for Predicting TVB-N and TVC. *J. Food Sci.* **2016**, *81*, M906–M912. [CrossRef] [PubMed]
5. Eglitis, R.I.; Purans, J.; Gabrusenoks, J.; Popov, A.I.; Jia, R. Comparative Ab Initio Calculations of ReO_3, $SrZrO_3$, $BaZrO_3$, $PbZrO_3$ and $CaZrO_3$ (001) Surfaces. *Crystals* **2020**, *10*, 745. [CrossRef]
6. Zhang, Y.J.; Chen, E.X. Comprehensive Monitoring System of Fresh Food Cold Chain Logistics. *AMM* **2014**, *602–605*, 2340–2343. [CrossRef]
7. Ezati, P.; Bang, Y.-J.; Rhim, J.-W. Preparation of a Shikonin-Based PH-Sensitive Color Indicator for Monitoring the Freshness of Fish and Pork. *Food Chem.* **2021**, *337*, 127995. [CrossRef] [PubMed]
8. Zhang, J.; Zou, X.; Zhai, X.; Huang, X.; Jiang, C.; Holmes, M. Preparation of an Intelligent PH Film Based on Biodegradable Polymers and Roselle Anthocyanins for Monitoring Pork Freshness. *Food Chem.* **2019**, *272*, 306–312. [CrossRef]
9. Sun, W.; Li, H.; Wang, H.; Xiao, S.; Wang, J.; Feng, L. Sensitivity Enhancement of PH Indicator and Its Application in the Evaluation of Fish Freshness. *Talanta* **2015**, *143*, 127–131. [CrossRef]
10. Kim, D.; Lee, S.; Lee, K.; Baek, S.; Seo, J. Development of a PH Indicator Composed of High Moisture-Absorbing Materials for Real-Time Monitoring of Chicken Breast Freshness. *Food Sci. Biotechnol.* **2017**, *26*, 37–42. [CrossRef]
11. Liu, J.; Wang, H.; Wang, P.; Guo, M.; Jiang, S.; Li, X.; Jiang, S. Films Based on κ-Carrageenan Incorporated with Curcumin for Freshness Monitoring. *Food Hydrocoll.* **2018**, *83*, 134–142. [CrossRef]
12. Tranchemontagne, D.J.; Mendoza-Cortés, J.L.; O'Keeffe, M.; Yaghi, O.M. Secondary Building Units, Nets and Bonding in the Chemistry of Metal–Organic Frameworks. *Chem. Soc. Rev.* **2009**, *38*, 1257. [CrossRef]
13. Perry IV, J.J.; Perman, J.A.; Zaworotko, M.J. Design and Synthesis of Metal–Organic Frameworks Using Metal–Organic Polyhedra as Supermolecular Building Blocks. *Chem. Soc. Rev.* **2009**, *38*, 1400. [CrossRef] [PubMed]
14. Ma, S. Gas Adsorption Applications of Porous Metal–Organic Frameworks. *Pure Appl. Chem.* **2009**, *81*, 2235–2251. [CrossRef]

15. Uzun, A.; Keskin, S. Site Characteristics in Metal Organic Frameworks for Gas Adsorption. *Prog. Surf. Sci.* **2014**, *891*, 56–79. [CrossRef]
16. Li, J.-R.; Kuppler, R.J.; Zhou, H.-C. Selective Gas Adsorption and Separation in Metal–Organic Frameworks. *Chem. Soc. Rev.* **2009**, *38*, 1477. [CrossRef]
17. Montoro, C.; Linares, F.; Quartapelle Procopio, E.; Senkovska, I.; Kaskel, S.; Galli, S.; Masciocchi, N.; Barea, E.; Navarro, J.A.R. Capture of Nerve Agents and Mustard Gas Analogues by Hydrophobic Robust MOF-5 Type Metal–Organic Frameworks. *J. Am. Chem. Soc.* **2011**, *133*, 11888–11891. [CrossRef]
18. Koh, C.S.L.; Lee, H.K.; Han, X.; Sim, H.Y.F.; Ling, X.Y. Plasmonic Nose: Integrating the MOF-Enabled Molecular Preconcentration Effect with a Plasmonic Array for Recognition of Molecular-Level Volatile Organic Compounds. *Chem. Commun.* **2018**, *54*, 2546–2549. [CrossRef]
19. Lustig, W.P.; Mukherjee, S.; Rudd, N.D.; Desai, A.V.; Li, J.; Ghosh, S.K. Metal–Organic Frameworks: Functional Luminescent and Photonic Materials for Sensing Applications. *Chem. Soc. Rev.* **2017**, *46*, 3242–3285. [CrossRef]
20. Raja Lakshmi, P.; Nanjan, P.; Kannan, S.; Shanmugaraju, S. Recent Advances in Luminescent Metal–Organic Frameworks (LMOFs) Based Fluorescent Sensors for Antibiotics. *Coord. Chem. Rev.* **2021**, *435*, 213793. [CrossRef]
21. Yassine, O.; Shekhah, O.; Assen, A.H.; Belmabkhout, Y.; Salama, K.N.; Eddaoudi, M. H_2S Sensors: Fumarate-Based Fcu-MOF Thin Film Grown on a Capacitive Interdigitated Electrode. *Angew. Chem. Int. Ed.* **2016**, *55*, 15879–15883. [CrossRef] [PubMed]
22. Dong, M.-J.; Zhao, M.; Ou, S.; Zou, C.; Wu, C.-D. A Luminescent Dye@MOF Platform: Emission Fingerprint Relationships of Volatile Organic Molecules. *Angew. Chem. Int. Ed.* **2014**, *53*, 1575–1579. [CrossRef] [PubMed]
23. Guo, L.; Liu, Y.; Kong, R.; Chen, G.; Wang, H.; Wang, X.; Xia, L.; Qu, F. Turn-on Fluorescence Detection of β-Glucuronidase Using RhB@MOF-5 as an Ultrasensitive Nanoprobe. *Sens. Actuators B Chem.* **2019**, *295*, 1–6. [CrossRef]
24. GB/T 5009. *228-2016;* China National Food Safety Standard: Method for Analysis of Hygienic Standard of Meat and Meat Products. National Health and Family Planning Commission of the People's Republic of China: Beijing, China, 2016.
25. Mantanus, J.; Ziémons, E.; Lebrun, P.; Rozet, E.; Klinkenberg, R.; Streel, B.; Evrard, B.; Hubert, Ph. Active Content Determination of Non-Coated Pharmaceutical Pellets by near Infrared Spectroscopy: Method Development, Validation and Reliability Evaluation. *Talanta* **2010**, *80*, 1750–1757. [CrossRef]
26. Liu, H.; Ji, Z.; Liu, X.; Shi, C.; Yang, X. Non-Destructive Determination of Chemical and Microbial Spoilage Indicators of Beef for Freshness Evaluation Using Front-Face Synchronous Fluorescence Spectroscopy. *Food Chem.* **2020**, *321*, 126628. [CrossRef]
27. Lee, L.C.; Liong, C.-Y.; Jemain, A.A. Partial Least Squares-Discriminant Analysis (PLS-DA) for Classification of High-Dimensional (HD) Data: A Review of Contemporary Practice Strategies and Knowledge Gaps. *Analyst* **2018**, *143*, 3526–3539. [CrossRef]
28. Liu, Y.; Ng, Z.; Khan, E.A.; Jeong, H.-K.; Ching, C.; Lai, Z. Synthesis of Continuous MOF-5 Membranes on Porous α-Alumina Substrates. *Microporous Mesoporous Mater.* **2009**, *118*, 296–301. [CrossRef]
29. Chen, X.; Jia, J.; Ma, H.; Wang, S.; Wang, X. Characterization of Rhodamine B Hydroxylamide as a Highly Selective and Sensitive Fluorescence Probe for Copper(II). *Anal. Chim. Acta* **2009**, *632*, 9–14. [CrossRef]
30. Albani, J.R. *Principles and Applications of Fluorescence Spectroscopy*, 1st ed.; Wiley: Hoboken, NJ, USA, 2007; ISBN 978-1-4051-3891-8.

biosensors

MDPI

Article

Online Inspection of Browning in Yali Pears Using Visible-Near Infrared Spectroscopy and Interpretable Spectrogram-Based CNN Modeling

Yong Hao [1,2,*], Xiyan Li [1], Chengxiang Zhang [1] and Zuxiang Lei [3]

[1] School of Mechatronics and Vehicle Engineering, East China Jiaotong University, Nanchang 330013, China
[2] Key Laboratory of Conveyance Equipment of the Ministry of Education, Nanchang 330013, China
[3] School of Civil Engineering and Architecture, East China Jiaotong University, Nanchang 330013, China
* Correspondence: haonm@163.com

Abstract: Browning is the most common physiological disease of Yali pears during storage. At the initial stage, browning only occurs in the tissues near the fruit core and cannot be detected from the appearance. The disease, if not identified and removed in time, will seriously undermine the quality and sale of the whole batch of fruit. Therefore, there is an urgent need to explore a method for early diagnosis of the browning in Yali pears. In order to realize the dynamic and online real-time detection of the browning in Yali pears, this paper conducted online discriminant analysis on healthy Yali pears and those with different degrees of browning using visible-near infrared (Vis-NIR) spectroscopy. The experimental results show that the prediction accuracy of the original spectrum combined with a 1D-CNN deep learning model reached 100% for the test sets of browned pears and healthy pears. Features extracted by the 1D-CNN method were converted into images by Gramian angular field (GAF) for PCA visual analysis, showing that deep learning had good performance in extracting features. In conclusion, Vis-NIR spectroscopy combined with the 1D-CNN discriminant model can realize online detection of browning in Yali pears.

Keywords: visible-near infrared spectroscopy; deep learning; online analysis; browned Yali pears

Citation: Hao, Y.; Li, X.; Zhang, C.; Lei, Z. Online Inspection of Browning in Yali Pears Using Visible-Near Infrared Spectroscopy and Interpretable Spectrogram-Based CNN Modeling. *Biosensors* **2023**, *13*, 203. https://doi.org/10.3390/bios13020203

Received: 7 December 2022
Revised: 18 January 2023
Accepted: 21 January 2023
Published: 29 January 2023

1. Introduction

Yali pears (*Pyrus bretschneideri* Rehd.) are also known as white pears [1]. With a regular and duck head-like shape, this kind of pear has thin skin, a small kernel, and juicy flesh. Yali pears are loved by consumers at home and abroad because they have the functions of relieving cough and thirst, and clearing heat and detoxification [2]. Yali pears can be easily affected by their environment during storage, especially by carbon dioxide (CO_2). When the CO_2 concentration is greater than 1%, the Yali pears will suffer from internal browning [3]. Browning is mainly related to the regulation of various genes and relevant enzymes in Yali pears [4], among which phenolic substances are the main substrates leading to browning. Under aerobic conditions, phenolic substances react with polyphenol oxidase to produce quinones [5], thereby accelerating the occurrence of browning. The browning of Yali pears starts from the inside of the fruit, so it is difficult to observe from the surface at the initial stage. Only when the browning worsens, it can be observed that the surface of Yali pears will become darker. Browning undermines the quality of Yali pears and makes them less tasty, thus leading to bad sales. Therefore, it is urgent to explore a nondestructive, environment friendly, and accurate technology to identify browning in Yali pears.

During post-harvest storage, Yali pears are prone to core and pulp browning and lose their commercial value. Therefore, huge economic losses are caused to producers and operators. The traditional method of browning discrimination is usually combined with experience to conduct destructive sampling inspection of internal components [6]. This will make fruit analysis difficult to achieve quickly, accurately, and non-destructively. It is not suitable for the actual situation of fruit production and sales at present. Visible-near infrared (Vis-NIR) spectroscopy is an indirect measurement method suitable for rapid online analysis. With its environment-friendly, non-destructive, and flexible integrated detection unit, Vis-NIR spectroscopy has been widely used in the detection of fruit components and defects. Sun et al. [7] adopted Vis-NIR spectroscopy to simultaneously measure online the browned core and soluble solids content (SSC) in Yali pears. They selected a total of 200 samples, including 73 pears with black-heart disease and 127 pears of good quality. The results showed that the classification accuracy of black-heart Yali pears by Vis-NIR spectroscopy was up to 98.3%. The percentages of SSC predictive precision were 97.8% and 99% within deviations of ± 0.5 and $\pm 1\%$, respectively. Qin et al. [6] conducted online and non-destructive detection of moldy-heart disease in apples through the miniature Vis-NIR spectrometer. After the 96 samples were optimized in terms of placement posture, the accuracy of partial least squares discriminant analysis (PLS-DA) reached 93.75%. By adopting Vis-NIR spectroscopy, Hao et al. [8] established an AdaBoost integrated model based on k nearest neighbors (kNN), naive Bayes classifier (NBC), and support vector machine (SVM) for 285 Yali samples. The results showed that the AdaBoost model combined with wavelet transform (WT) had the highest discrimination accuracy (92.63%) of black-heart Yali online detection. Cruz et al. [9] simulated the randomness of fruit positions during spectral acquisition by randomly sampling on the four sides of 1002 'Rocha' pears, so as to figure out the optimal combination of pretreatment method and discrimination method. The results showed that PLS-DA combined with the original spectrum of Yali pears browning classification mode presented the best performance, and the classification accuracy reached 83%. The above results indicated that the research on the discriminant model of browning of Yali pears has basically stayed on machine learning in recent years. These methods are applicable when the amounts of samples are relatively small, and the accuracy requirement is not high. There were few studies on the discriminant of Yali pears by the deep learning model combined with the spectral analysis method. Vis-NIR as an environment-friendly and nondestructive detection technique can be employed to identify browning in Yali pears. When the number of samples is limited, the discrimination accuracy will vary with different identification methods.

In recent years, convolutional neural network (CNN) has been widely and successfully used in image recognition, natural language, video, and other fields. By combining NIR technology with one-dimensional convolutional neural network (1D-CNN), Li et al. [10] developed a data-driven model to estimate the content of organic matter in *Huangshan Maofeng* tea. The results showed that the key features of NIR spectroscopy were successfully extracted, thus providing a new and effective NIR analysis strategy for food analysis. Wu et al. [11] established a quantitative analysis model to detect olive oil content in a corn oil–olive oil mixture by combining Raman spectroscopy with 1D-CNN. The results showed that the 1D-CNN model based on 315 extended average Raman spectra could quantitatively detect the content of olive oil, in which the predictive determination coefficient (R_P^2) and the root mean square error of prediction (RMSEP) were 0.9908 and 0.7183, respectively. Chen et al. [12] established a CNN calibration model for the NIR quantitative determination of water pollution and verified the applicability of shallow convolutional network modeling architecture in feature extraction of one-dimensional spectral data. Rong et al. [13] proposed a detection method based on the principle of deep learning. Through the construction of a 1D-CNN model, the Vis-NIR spectral database containing a total of 500 samples of five kinds of peaches was established for identifying peach varieties. The results showed that the accuracy of the deep learning model reached 100% in the validation data set and 94.4% in the test data set. As suggested by multiple studies, when the number of samples for analysis met certain requirements, CNN combined with Vis-NIR can be applied for the identification of material variety and component analysis. In addition, for the NIR spectral analysis based on scientific research instruments or at a static state, the CNN method can obtain better analytical accuracy, because the spectral response has better wavelength accuracy and less external noise interference.

Vis-NIR online analysis technology cannot be successfully applied without a stable, accurate, and concise modeling algorithm. The effectiveness of the algorithm not only depends on the information extraction ability of the algorithm itself, but also the pretreatment of spectra and effective variable screening before modeling. However, a large number of discriminative models need to be established for the continuous selection of multiple spectral pretreatment methods and variable screening methods, and it is hard to maintain the consistency of optimal variables obtained by different variable screening methods, which affects the stability of the model. For this reason, in order to ensure the accuracy of the model while reducing the complexity and differences in the selection of pretreatment methods and spectral variable screening methods, this paper intends to adopt a deep learning method to simultaneously realize the automatic extraction of spectral feature variables and model construction. In other words, the depth features of Vis-NIR of Yali pears are extracted and used for modeling by means of a 1D-CNN algorithm, so that the interference of noise on the feature variables selection can be avoided and the stability of the model is thus improved. This study can provide reference for rapid online analysis of Yali pear quality in an accurate and reliable manner based on deep learning.

2. Materials and Methods

2.1. Sample Preparation

The Yali pear samples in this paper were collected from an orchard in Hebei Province, China, and then were refrigerated and transported to the laboratory immediately after harvesting. In order to reduce the influence of temperature on spectrum acquisition [14] and the damage caused by rapid cooling to the fruit, these samples were stored for 24 h under constant temperature before experiments. The sample storage temperature was 1 °C (±0.5), and the humidity was 85–95%. The diameter of Yali pears selected in the experiment was about 75 mm, and they had no obvious defects and mechanical damage on the surface. They were cleaned and numbered. A total of 495 samples were collected, including 256 healthy pears and 239 browned pears. The Kennard–Stone algorithm was used to divide healthy and browned pear samples into calibration sets and test sets in the ratio of 7:3.

2.2. Vis-NIR Spectroscopy Acquisition

The Vis-NIR spectra of Yali pears were collected by an Ocean Optics INC, QE65Pro (Dunedin, FL, USA) high-precision spectrometer with a wavelength range of 361–1165 nm. The spectral resolution of ca. 0.8 nm and the signal-to-noise ratio (SNR) of 1000 were employed. Before spectral acquisition, the spectrometer was preheated for 30 min to obtain stable light source energy. Then, a PTFE ball was used to conduct the qualified calibration on the spectrometer and the sorting device. Figure 1 shows the dynamic online acquisition device of Yali pears. The online detection device, which consists of spectrometer, light source, conveyor belt, optical fiber (model QP1000-2-VIS-NIR, Ocean Optics, Dunedin, FL, USA), and computer and so on, includes three modules (transmission module, spectral acquisition module, and control module).

The spectrometer integration time was set to 80 ms, and the conveyor belt can transmit six samples per second. The optical fiber probe was located below the tray, at a distance of about 12 mm from the conveyor belt. The light source, which consists of ten 100-watt halogen tungsten lamps (Osram), was located above the sample and arranged along the concentrating coil at an angle of 45 degrees. The halogen tungsten lamps were arranged in an equidistant arrangement of five lamps on both sides of the Yali pear samples, as shown in Figure 1C.

Yali pears were manually placed on the device for spectral acquisition, and the acquisition process is as follows. First, as shown in Figure 1C, the pear samples were placed in a way that the connecting line direction between the pears' stalk and the pears' pedicel was always perpendicular to the running direction of the conveyor belt. When the pears were transported to the spectral acquisition station by a conveyor belt, the arc-shaped equidistant light source on both sides could evenly penetrate the pears into the detector probe. Second, the spectrometer was triggered by signals sent by the system encoder, which acquired the Vis-NIR spectrum of Yali pear samples. The collected transmission spectra were automatically stored on the computer. Each sample was scanned five times in order to ensure the consistency of spectral collection, and the average spectrum was used as the final spectrum of the sample. Each sample spectrum had 1044 variables in the wavelength range of 361–1165 nm. As the signal-to-noise ratios at both ends of the detector were low, 1024 variables in the wavelength range of 387–1165 nm were thus reserved for subsequent experiments. The whole spectral acquisition process was completed in the designed shield that effectively avoided the influence of external stray light.

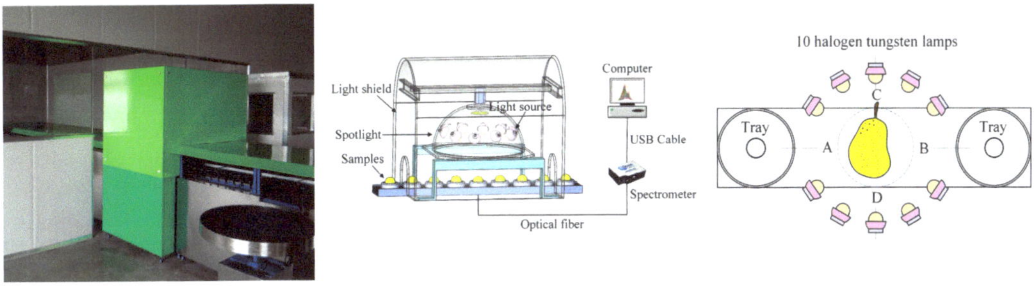

(A) (B) (C)

Figure 1. Dynamic online acquisition device of Yali pears: (**A**) device diagram; (**B**) core detection unit; (**C**) schematic diagram of arrangement and distribution of tungsten halogen lamps.

2.3. Evaluation of Browning in Yali Pears

The damage information of browning in Yali pears was evaluated by the destructive method. After spectral acquisition, each pear sample was cut along the direction of the CD line in Figure 1C, and three experts with years of experience in pear planting determined whether the inside of these pears had browned. The evaluation criteria are as follows: if there were no abnormal changes of cell tissue near the pears' core, then it was a healthy pear; while if tissues around the core were browned, then it was classified as a browned pear. The picture of flesh inside browning pears is shown in Figure 2, and it can be clearly seen in the figure that there are varying degrees of browning near the core.

Figure 2. Photo of the internal pulp of pears with different degrees of browning.

2.4. Spectral Pretreatment and Variable Selection

The original Vis-NIR data contain not only feature information, but also many noise and other interference factors, such as the temperature and humidity of the environment during the spectral acquisition process, the location of the samples, and the light source stability. In order to achieve modeling spectra with better quality and enhance the features with stronger correlation in the spectra, the traditional pretreatment methods such as first-derivative (SG 1st−Der) [15], multiplicative scatter correction (MSC), standard normal variable transformation (SNV) [16], moving average smoothing (MAS), and wavelet transform (WT) [17] were used in this experiment for modeling. In this case, problems of baseline shift, scattering, and high-frequency noises of the original spectra can be removed.

The spectra of Yali pear samples contain many uninformative variables which are characterized by a highly overlapping spectral band. In order to eliminate redundant and uninformative spectral variables, Monte Carlo uninformative variable elimination (MCUVE) was used in this paper to select effective wavelengths, which can significantly reduce the number of feature wavelengths and the complexity of the model, as well as improve the generalization ability and robustness of the model [18]. In MCUVE, the contributions of wavelength variables in high dimensional spectral data were evaluated by making full use of the intrinsic correlation among samples. The contribution value of each spectrum, which served as an indicator of stability, was sorted to establish a series of partial least squares regression (PLSR) models. The importance of each variable was measured by its stability value, and the threshold value of variable optimization was finally determined by the minimum root mean square error of cross-validation.

2.5. Construction Method of Discriminant Model

2.5.1. PLS-DA Model

PLS-DA, or partial least squares discriminant analysis, is a discriminant analysis method used to deal with classification and discrimination problems in high-dimensional data analysis. It can effectively distinguish the observed values between classes and find the influencing variables that lead to the differences between classes [19]. In addition, PLS-DA can reduce the influence of multicollinearity among variables. A simple linear model can be used to describe the relationship between variables and responses, as shown in Equation (1).

$$Y = b_0 + b_1X_1 + b_2X_2 + \ldots + b_nX_n \tag{1}$$

where b_0 refers to intercept, b_i is the regression coefficient of the sample, and X_i is the ith spectral response value corresponding to the wavelength.

2.5.2. SVM Model

The basic principle the support vector machines (SVM) algorithm is to figure out the separation hyperplane that can correctly partition the training data sets and maximize the geometric interval and map the input vectors to a high-dimensional feature space using a specific transfer kernel function [20]. The construction of a linear surface for feature space endows the SVM network with higher generalization ability. The optimization objectives of SVM are presented as follows.

$$\begin{cases} \min\limits_{\omega,b,\xi_i} \frac{1}{2}\|\omega\|^2 + C\sum\limits_{i=1}^{n}\xi_i \\ \text{s.t.}\begin{cases} y_i(\omega \cdot x_i + b) \geq 1 - \xi_i \\ \xi_i \geq 0,\ i = 1, 2, \ldots, N \end{cases} \end{cases} \tag{2}$$

where ω is the normal vector that determines the direction of the hyperplane; b is the displacement term which determines the distance between the hyperplane and the original point; ζ_i is the relaxation variable which represents the degree to which the sample does not satisfy the constraint; C is the penalty factor, and the greater its value is, the larger value indicates a greater penalty for classification; n is the number of samples; x_i is the supporting vector of training sample; y_i is the category of the corresponding sample.

2.5.3. 1D-CNN Model

Integrating feature learning and classification, the convolutional neural network (CNN) is a feed-forward neural network with convolution computation and deep structure. By mapping simple data features to complex high-dimensional space by convolution and pooling, CNN can obtain the accuracy of classification. Studies have shown that the convolution layer can extract important features with more useful information while removing noises and unimportant features, which is a more robust algorithm for feature extraction than the feature selection method [21]. The working principle of 1D-CNN is shown in Figure 3, where the dimension of the convolution kernel is 3×1 and the moving step is 1. The data features are extracted through distributively moving the convolution kernel.

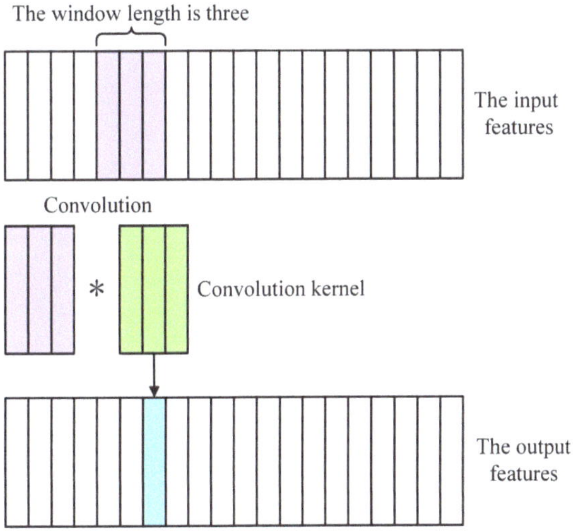

Figure 3. Schematic diagram of 1D-CNN.

The structure of the 1D-CNN network consists of the convolution layer, pooling layer, and full connection layer. The operation of the convolution layer is equivalent to the process of pretreatment and feature extraction in traditional machine learning [22]. Before convolution, the size of the convolution kernel and the number of filters need to be set and these two parameters are obtained through multiple experiments. The maximum pooling was selected as the pooling layer, which can reduce the dimension of feature mapping without losing important information, thus cutting computation amount and time of network training. The fully connected layer acted as the classifier in CNN and optimizes the network as the feature combination layer. In order to prevent over-fitting in training data, a dropout layer was added to the network structure to suppress over-fitting by removing some neurons. The flow chart of the network framework of the 1D-CNN model is shown in Figure 4.

All models and chemometric procedures used throughout the work were implemented based on Python 3.8.

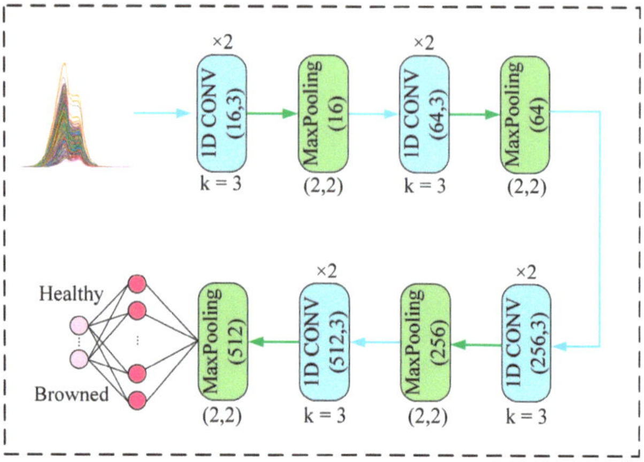

Figure 4. The flow chart of the network framework of the 1D-CNN model.

2.6. Evaluation of Models

Three indexes including the overall accurate identification rate (*Accuracy*), accurate identification rate of healthy pears (*RH*), and accurate identification rate of browned pears (*RB*) were adopted to evaluate the online discriminative model of browning in Yali pears. *Accuracy* refers to the rate of correct identification made by the classifier for all samples; *RH* refers to the percentage of correct identification made by the classifier for all healthy samples; *RB* refers to the percentage of correct identification made by the classifier for all browned samples. The greater the values of these three indexes are, the higher the rate of correct classification is. *Accuracy*, *RH*, and *RB* can be calculated according to the following equations.

$$Accuracy = \left(1 - \frac{He + Be}{H + B}\right) \times 100\% \quad (3)$$

$$RH = \left(1 - \frac{He}{H}\right) \times 100\% \quad (4)$$

$$RB = \left(1 - \frac{\text{Be}}{B}\right) \times 100\% \quad (5)$$

where *H* refers to the total number of healthy samples, and *He* refers to the number of healthy samples being mistakenly classified. *B* is the total number of browned samples, and *Be* is the number of browned samples being mistakenly classified.

3. Results and Discussion

3.1. Vis-NIR Spectral Analysis of Yali Pears

Figure 5 shows the original Vis-NIR spectra of healthy and browning Yali pears, as well as the averaged spectra. It can be seen from Figure 5A that the spectra of healthy pears and browning pears overlap so seriously that it is impossible to directly distinguish from the spectral graph whether Yali pears were browned or not. It can be seen from Figure 5B that the average spectrum of healthy pears is higher than that of browned pears on the whole, and the band of 600–800 nm is the most obvious, which may be due to the strong absorption of transmitted light by browning tissue inside the fruit. As shown in the spectral graph, there are two absorption peaks at approximately 700 and 800 nm. In addition, the absorption peak at around 700 nm may result from the stretching and contraction of the fourth overtone of the C-H functional group, while that at around 800 nm may be related to the stretching and contraction of the third overtone of the N-H functional group [23].

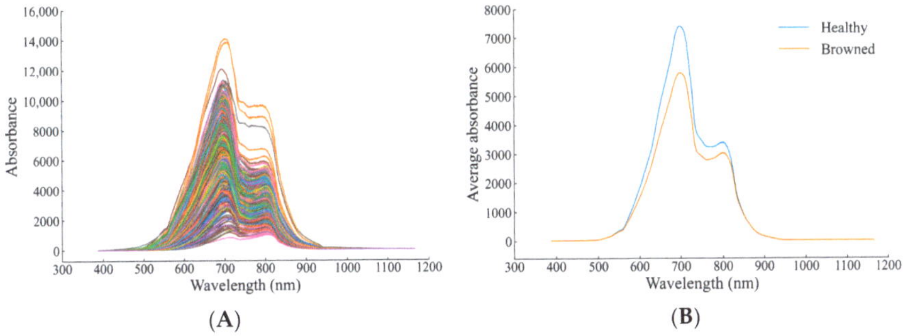

Figure 5. Vis-NIR spectra of healthy and browned pears: (**A**) raw spectra; (**B**) average spectra.

3.2. PCA Analysis on Health Status of Yali Pears with Different Pretreatment Methods

In the experiment, principal component analysis (PCA) was applied to analyze the spectral spatial distribution of two kinds of Yali pear samples pretreated by different methods. The core idea of PCA is dimensionality reduction, and its working principle is to transform a group of variables that may have correlation into a set of linearly uncorrelated variables, namely principal components, through orthogonal transformation. The data in the new subspace defined by the principal component are usually easier to interpret [24].

The data set contains 256 healthy pears and 239 browned pears. PCA analysis was conducted on the original spectrum and on the spectra separately pretreated by SG 1st−Der, MSC, SNV, MAS, and WT. The first three principal components were retained for visualization. The results are shown in Figure 6. It can be observed that the spatial distribution of spectra pretreated by MAS and WT were basically consistent with that of the original spectrum. The spectral points of browned pears and healthy pears were so crossed that they cannot be distinguished, indicating that these two pretreatment methods failed to significantly improve the spectral features of data used in this experiment. Compared with the original spectrum, pretreatment methods of SG 1st−Der, MSC, and SNV have shown better performance. The spectral points of samples pretreated by these three methods presented a trend of classification separation in terms of spatial distribution. The cumulative contribution rates of the first three principal components of SG 1st−Der, MSC, and SNV were 96.4%, 99.6%, and 99.3%, respectively. The cumulative contribution rates of MSC and SNV were both over 99%, indicating that both methods were applicable to the experimental data. In order to further explore the effectiveness of the method, PLS-DA and SVM models were established combined with different pretreated methods. The results of the discrimination are shown in Table 1. It can be seen from the table that the discrimination accuracy of SNV is higher than that of MSC. Therefore, the SNV pretreatment method was finally selected for model optimization in this experiment.

Table 1. The discrimination results of PLS-DA and SVM models' training sets under different preprocessing methods.

Model	Pretreatment	Accuracy (%)	LVs	Model	Pretreatment	Accuracy (%)
PLS-DA	Raw	82.66	4	SVM	Raw	96.53
	SG 1st−Der	95.09	5		SG 1st−Der	100
	MSC	95.66	8		MSC	100
	SNV	97.11	9		SNV	100
	MAS	84.97	5		MAS	96.25
	WT	85.26	4		WT	95.66

Polynomial order of SG 1st−Der is three, and the number of smoothing points is 15. The number of smoothing points in MAS is 15. Daubechies 8 wavelet is selected in WT.

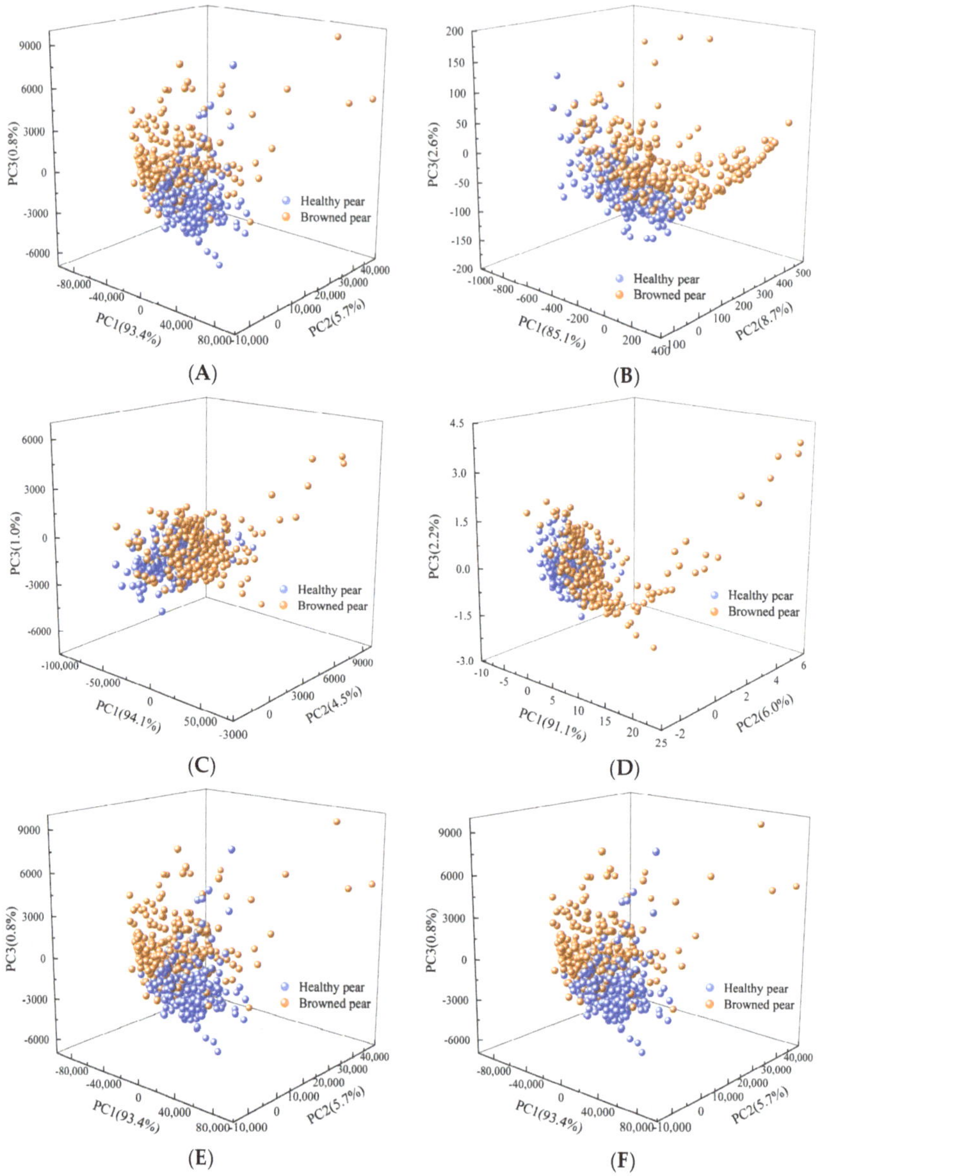

Figure 6. The spatial distributions of the first three principal components of healthy pear and browned pear samples pretreated by different methods: **(A)** raw; **(B)** SG 1st−Der; **(C)** MSC; **(D)** SNV; **(E)** MAS; **(F)** WT.

3.3. Robust Variable Selection Based on MCUVE Method

When the spectrum contains a large number of invalid variables, it will undermine the quality of modeling to a certain extent. In order to further simplify the model and improve the accuracy and interpretability of the model, the spectral wavelength was optimized by the MCUVE method after SNV pretreatment. As variables were selected in a random way, each model was selected 10 times. The optimized results are shown in Figure 7. Figure 7A shows the corresponding relationship between the variable distribution of the 10-time wavelength selections and the corresponding spectrum. It can be seen that the wavelength distribution of each optimal selection is not completely consistent, but most of the variables are concentrated at the absorption peak of the sample. Figure 7B shows the number of variables retained after 10 runs of MCUVE, and it can be seen that the number is within the range of 224 to 324. Although the distribution of the selected variables is relatively concentrated, the number of retained variables is different. When the distribution of the modeling samples is not representative or the number is small, although the MCUVE method can improve the accuracy of the model, the number of wavelengths selected for multiple times is unstable, which affects the robustness of the model.

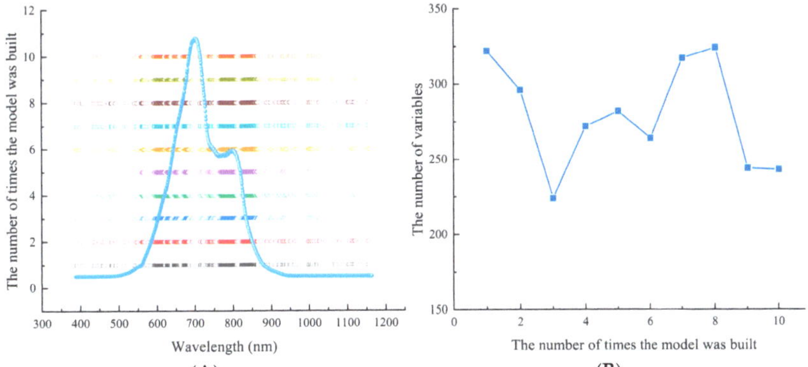

Figure 7. The results of variables selected after 10 runs of MCUVE method: (**A**) variable distribution; (**B**) the number of variables retained.

3.4. Evaluation Model for Health Status of Yali Pears Based on PLS-DA Method

PLS, which contains least squares regression analysis and discriminant analysis, is suitable for analyzing cases where the differences between groups are small while differences within groups are large. Unlike PCA, PLS-DA is a supervised discriminant analysis statistical method, which is widely used in the discrimination of Vis-NIR spectral data analysis. The core idea of PLS-DA is to obtain the optimal number of factors (LVs) in the calibration set through Monte Carlo cross-validation combined with an F-test. In this experiment, 10-fold cross-validation was employed to determine LVs. Furthermore, the results show that the root mean square error of the calibration set was the smallest when the LVs were 9, indicating that the model presented the best classification performance when the LVs were 9.

The model was established based on the calibration set, and then its performance was verified by the sample test set. The modeling results of the full spectrum, the pretreated spectrum, and the optimally selected spectrum (the best model in the 10 selections is chosen, whose number of variables is 264) are shown in Table 2. It can be seen that after spectral pretreatment and variable selection, the discrimination accuracy of discriminating quality of Yali pears had been improved. The optimal model of PLS-DA is SNV-MCUVE-PLS-DA, and accuracy, RH, and RB of its test set are 97.32%, 100%, and 92.16%, respectively.

Table 2. The results of PLS-DA discriminative models for pear quality based on different modeling variables.

Model	Method	Accuracy (%)	RH (%)	RB (%)
PLS-DA	Raw	80.54	88.64	68.85
	SNV	95.97	98.98	90.20
	SNV-MCUVE	97.32	100	92.16

RH: accurate identification rate of healthy pears; RB: accurate identification rate of browned pears.

3.5. Evaluation Model for Health Status of Yali Pears Based on SVM Method

SVM is a class of generalized linear classifiers that perform binary classification of data based on the supervised learning method. SVM mainly maximizes the interval between different categories by finding the division hyperplane with the largest interval. It is a class of generalized linear classifiers for binary classification of data in a supervised learning manner. The selection of SVM hyperparameters (kernel function, penalty factor C, and kernel function parameter g) is determined by an optimizing algorithm through grid search so as to obtain the best calibration model. Grid search is an exhaustive search method, which uses cross-validation to optimize the estimation function and finally solves the optimal parameters. In this experiment, grid search determines that linear is the optimal kernel function, the value of the optimal penalty factor C is 100, and the value of the optimal kernel function parameter g is 0.001.

It can be seen from Table 3 that the optimal SVM model is SNV-MCUVE-SVM, and accuracy, RH, and RB of its test set are 98.66%, 100%, and 96.08%, respectively; moreover, the modeling results of the SNV-SVM model are exactly the same as those of the SNV-MCUVE-SVM model. There are only 317 variables involved in the modeling of the SNV-MCUVE-SVM model, which greatly reduces the complexity of the model. Therefore, the optimal SVM model is the SNV-MCUVE-SVM model in which spectra are pretreated by SNV and variables are optimally selected by MCUVE.

Table 3. The results of SVM discriminative models for pear quality based on different modeling variables.

Model	Method	Accuracy (%)	RH (%)	RB (%)
SVM	Raw	95.97	97.73	93.44
	SNV	98.66	100	96.08
	SNV-MCUVE	98.66	100	96.08

3.6. Evaluation Model for Health Status of Yali Pears Based on 1D-CNN Method

The above research shows that although traditional methods can classify the browning pears, the process is cumbersome. A detection model with good performance requires rich theoretical knowledge and practical experience of the modeler. For this reason, it is necessary to figure out the effect of the deep feature extraction method on discriminating browned pears and healthy pears.

This paper proposed building a 1D-CNN model based on one-dimensional Vis-NIR spectral data in a bid to determine more accurately whether Yali pears are browned or not. The network was composed of an input layer, convolutional layer, pooling layer, and full connection layer. These layers were mainly used for feature extraction and classification. Among them, the convolutional layer and the pooling layer were used to extract the features of the data, while the fully connected layer was used to map the previously extracted features to the output space for subsequent classification. The activation functions of the convolutional layer and the full connection layer were the Relu function and the Sigmoid function, respectively. The dropout rate was set to 0.3. The Adam optimizer was adopted to optimize the network, the learning rate of the training model was set to 0.0001, the batch size was set to 64, and the number of iterations (epochs) was set to 500. The output feature value of the full connection layer which can combine features was set to 400.

The spectral data of the calibration set was read into the initialization network for iterative training, while the spectral data of the test set was used to evaluate the accuracy of the deep learning model. The deep learning model was randomly run 10 times in order to test its robustness, with the training samples and test samples unchanged. The modeling results of 10 runs are shown in Table 4. It can be seen that the correct discrimination rate of the 1D-CNN discriminant model has the lowest rate of 96.64% and the highest rate of 100%. The difference between the two is only 3.36%, indicating that the model is relatively robust.

Table 4. The results of 10 parallel runs of the 1D-CNN discriminant model for pear quality.

Modeling Method	The Model Number	Accuracy (%)	RH (%)	RB (%)
1D-CNN	1	99.33	100	98.36
	2	98.66	98.86	98.36
	3	96.64	98.86	93.44
	4	99.33	100	98.36
	5	98.66	98.86	98.36
	6	97.32	95.45	100
	7	98.66	100	96.72
	8	100	100	100
	9	98.66	98.86	98.36
	10	99.33	100	98.36

3.7. Establishment and Error Analysis of the Optimal Evaluation Model for Health Status of Yali Pears

PLS-DA, SVM, and 1D-CNN discriminant methods were separately used to establish online models for identifying healthy pears and browned pears. These models were then adopted to qualitatively discriminate healthy pears and browned pears that were not involved in the modeling. In order to optimize the model and improve the discriminant performance of the model, the spectral preprocessing and variable selection were performed before PLS-DA and SVM modeling. As errors of spectral rotation and shift were eliminated, the model became less complicated. After the network was completely designed, the 1D-CNN discriminant model can be directly constructed based on the original spectrum of Yali pears, because 1D-CNN had excellent performance in integrating pretreatment and extracting features.

The discrimination results of training sets and test sets in 10 PLS-DA, SVM, and 1D-CNN models are shown in Figure 8. It can be seen that 1D-CNN has better discriminant performance than PLS-DA and SVM, as evidenced by better modeling and testing accuracy of all its discriminant models. Moreover, in the 1D-CNN method, there is no requirement for preprocessing and variable selection of sample spectra. The model with the highest discrimination accuracy in 10 times of 1D-CNN modeling is selected as the final model, and the discrimination results are shown in Table 5.

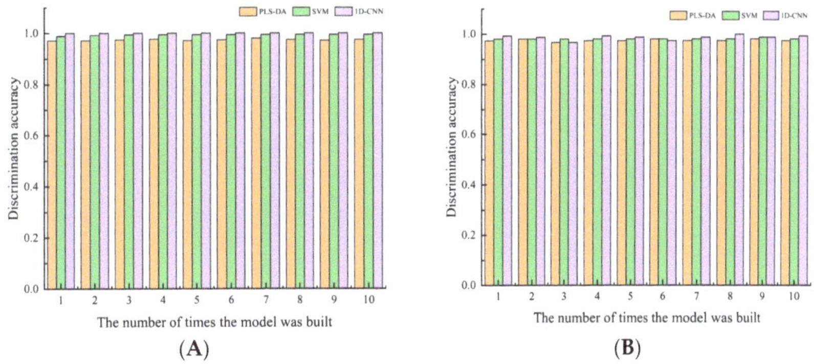

Figure 8. Discrimination accuracy of training sets and test sets in 10 PLS-DA, SVM, and 1D-CNN models: (**A**) training sets; (**B**) test sets.

Table 5. Statistics of test results of SVM, PLS-DA, and 1D-CNN models.

Samples in Test Set	SNV-MCUVE-PLS-DA		SNV-MCUVE-SVM		1DCNN	
	Healthy	Browned	Healthy	Browned	Healthy	Browned
Discrimination	98	51	98	51	88	61
Misclassification	0	4	0	2	0	0
RH (%)	100		100		100	
RB (%)	92.16		96.08		100	
Accuracy (%)	97.32		98.66		100	
Prediction Time (s)	0.0113		0.0099		0.0256	

It can be seen from the table that the PLS-DA, SVM, and 1D-CNN models all can correctly identify healthy pears, but there are also cases in which browned pears are misjudged as healthy pears. From the perspective of the data in experiment, the overall discrimination performance of the 1D-CNN model is better than that of PLS-DA and SVM models, as the discrimination accuracy of its test set was as high as 100%. Furthermore, the calculation time of the 1D-CNN model in predicting the test set samples was about 0.0256 s, which met the requirements of online classification. Compared with the traditional models, the 1D-CNN model can automatically identify important features and has better classification performance. Furthermore, this model is so simple to operate that even those without much basic knowledge of modeling can use it properly. In summary, the 1D-CNN model is the optimal model for discriminating browned pears.

3.8. Deep Feature Analysis on Vis-NIR Spectra of Yali Pears

It can be seen from the above studies that the deep learning model has excellent performance in extracting features of Vis-NIR spectral analysis for Yali pears. In order to carry out a further expression and analysis of the spectral features extracted by the 1D-CNN method, the Gramian angular field (GAF) was used to transform the spectral data into graphs. The GAF has been used for visual expression of one-dimensional time series signals and the good classification results have been obtained [25,26]. The main advantage of the GAF is that features of the original one-dimensional spectral signal of the image during encoding are maintained and bidirectionally mapped to the two-dimensional image [27]. The feature data extracted by 1D-CNN are converted into a 20×20 image by the GAF. The GAF encoding process of the Yali pears' spectral data is shown in Figure 9. Figure 9A,B show the results of the GAF transformation of healthy pears and browned pears, respectively.

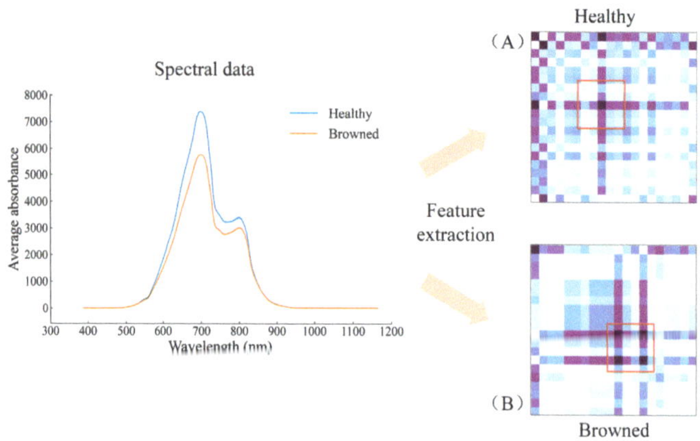

Figure 9. GAF coding process of spectral data of Yali pears.

Figure 10 shows five pictures randomly selected from the two types of pear samples. Among them, Figure 10A,B correspond to the GAF expression of healthy and browned pears in the test set, respectively. It can be seen from Figure 10 that the model can adaptively learn the separable features from the original spectral information. The GAF graphs of healthy pears are similar (all in the shape of '+'); the GAF graphs of browned pears are slightly different due to the different browning degrees of tissue cells near the fruit core, but most of the graphs are shown in Figure 10B (in the shape of '#'). It can be distinguished that the GAF graphs of the two types of samples are quite different, which further confirms that the deep learning model works excellently in feature extraction and classification. Meanwhile, the local and global features of Vis-NIR which can be easily classified and distinguished, if integrated by the feature combination layer (the full connection layer), can save calculation time while retaining important information.

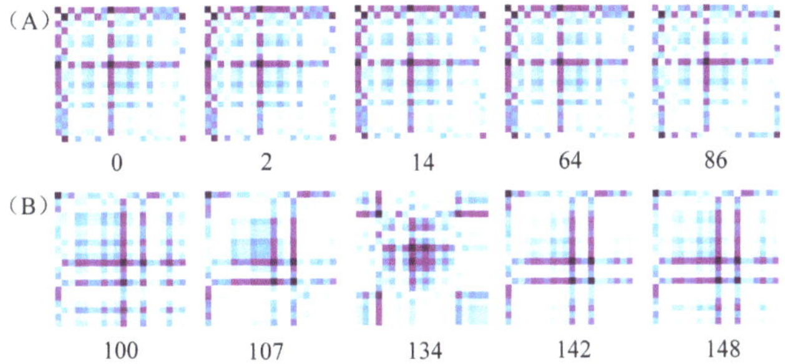

Figure 10. GAF graphs of spectra of some pear samples: (**A**) healthy pears; (**B**) browned pears.

The GAF features were visualized by the PCA method. The two-dimensional feature distributions of the original spectral data and of features extracted by the 1D-CNN are shown in Figure 11A,B. It can be seen from Figure 11A that the spatial distribution of the original spectral data is relatively chaotic; the spectral points of the two types of samples overlap each other, so the feature distinction is not obvious. It can be seen from Figure 11B that the data features of 1D-CNN after dimensionality reduction by PCA have better aggregation and feature distinction.

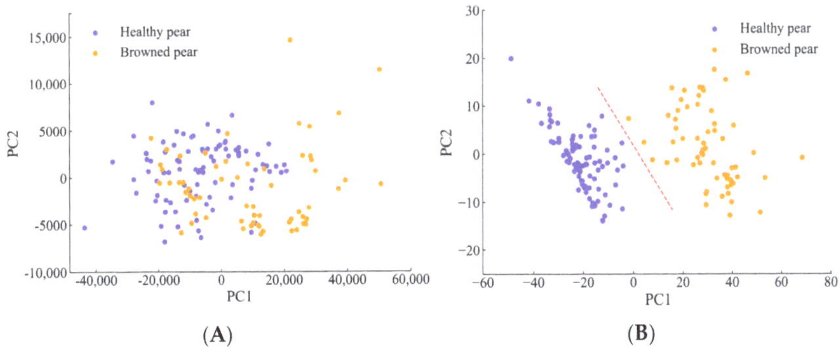

Figure 11. Two−dimensional spatial distribution of pear samples: (**A**) raw spectral data; (**B**) feature data extracted by 1D−CNN method.

4. Conclusions

Based on the in-depth study of 1D-CNN, this paper proposed a device and method for online detection of Yali pear browning using Vis-NIR spectroscopy combined with deep learning. SVM, PLS-DA, and 1D-CNN models were established to discriminate the browned pears. The original spectrum and five spectra pretreated by SVM and PLS-DA methods were used to establish and optimize the discriminative model for browning in Yali pears. Moreover, the MCUVE method was adopted to reduce the complexity of the model. The experimental results show that the energy spectra processed by SNV and MCUVE were adopted to establish the PLS-DA discriminant model, and the discrimination accuracy rate was 97.32%; the energy spectra processed by SNV and MCUVE were used to establish the SVM discriminative model, and the discrimination accuracy rate was 98.66%. Without cumbersome pretreatment and variable screening, the 1D-CNN discriminative model has shown the optimal performance, and the discrimination accuracy rate was as high as 100%. In conclusion, Vis-NIR spectroscopy combined with the 1D-CNN discriminant model can realize online detection of browning in Yali pears.

Author Contributions: Y.H.: Conceptualization, Methodology, Writing—review and editing, Funding acquisition. X.L.: Writing—original draft and editing, Software, Visualization, Resources. C.Z.: Investigation, Resources. Z.L.: Investigation, Resources. All authors have read and agreed to the published version of the manuscript.

Funding: This research was funded by the National Natural Science Foundation of China (Grant No. 31960497) and Jiangxi Provincial Natural Science Foundation of China (Grant No. 20212BAB204009 & 20202ACB211002).

Institutional Review Board Statement: Not applicable.

Informed Consent Statement: Not applicable.

Data Availability Statement: The data used to support the study findings are available from the corresponding authors upon request.

Conflicts of Interest: The authors declare no conflict of interest.

References

1. Han, D.; Tu, R.; Lu, C.; Liu, X.; Wen, Z. Nondestructive detection of brown core in the Chinese pear 'Yali' by transmission visible–NIR spectroscopy. *Food Control.* **2006**, *17*, 604–608. [CrossRef]
2. Guang, Y.; Li, W.; Wen, S.; Chang, S. Research Progress on Functional Ingredients and Food Development of Ya Pear. *Farm Prod. Process.* **2021**, *06*, 59–62. [CrossRef]
3. Du, Y.; Wang, W.; Jia, X.; Le, W.; Wei, J.; Guan, J. Development Status and Proposal of Pear Storage Industry of Hebei Province. *Storage Process* **2017**, *17*, 1–6. [CrossRef]
4. Wang, Z.; Zhang, Y.; Li, Y.; Li, L.; You, L.; Li, X.; Jin, Z.; Yan, S. Relationship Between LAC Gene Expression and Core Browning of Yali Pear. *Sci. Agric. Sin.* **2020**, *53*, 5073–5080. [CrossRef]
5. Ren, Z.; Cheng, Y.; Guan, Y.; Zhang, Z.; Guan, J. Effects of Mechanical Damage on Browning in 'Yali' Pear. Food Science and Technology. *Food Sci. Technol.* **2021**, *46*, 35–41. [CrossRef]
6. Qin, K.; Chen, G.; Zhang, J.; Fu, X. Optimization of Fruit Pose and Modeling Method for Online Spectral Detection of Apple Moldy Core. *Spectrosc. Spectral Anal.* **2021**, *41*, 3405–3410. Available online: http://www.gpxygpfx.com/CN/10.3964/j.issn.1000-0593(2021)11-3405-06 (accessed on 27 June 2022).
7. Sun, X.; Liu, Y.; Li, Y.; Wu, M.; Zhu, D. Simultaneous measurement of brown core and soluble solids content in pear by on-line visible and near infrared spectroscopy. *Postharvest Biol. Technol.* **2016**, *116*, 80–87. [CrossRef]
8. Hao, Y.; Wang, Q.; Zhang, S. Study on Online Detection Method of "Yali" Pear Black Heart Disease Based on Vis-Near Infrared Spectroscopy and AdaBoost Integrated Model. *Spectrosc. Spectral Anal.* **2021**, *41*, 2764–2769. Available online: http://www.gpxygpfx.com/CN/10.3964/j.issn.1000-0593(2021)09-2764-06 (accessed on 27 June 2022).
9. Cruz, S.; Guerra, R.; Brazio, A.; Cavaco, A.M.; Antunes, D.; Passos, D. Nondestructive simultaneous prediction of internal browning disorder and quality attributes in 'Rocha' pear (*Pyrus communis* L.) using VIS-NIR spectroscopy. *Postharvest Biol. Technol.* **2021**, *179*, 111562. [CrossRef]
10. Li, M.; Pan, T.; Bai, Y.; Chen, Q. Development of a calibration model for near infrared spectroscopy using a convolutional neural network. *J. Near Infrared Spectrosc.* **2022**, *30*, 89–96. [CrossRef]
11. Wu, X.; Gao, S.; Niu, Y.; Zhao, Z.; Ma, R.; Liu, H.; Zhang, Y. Quantitative analysis of blended corn-olive oil based on Raman spectroscopy and one-dimensional convolutional neural network. *Food Chem.* **2022**, *385*, 132655. [CrossRef] [PubMed]
12. Chen, H.; Chen, A.; Xu, L.; Xie, H.; Qiao, H.; Lin, Q.; Cai, K. A deep learning CNN architecture applied in smart near-infrared analysis of water pollution for agricultural irrigation resources. *Agric. Water Manag.* **2020**, *240*, 106303. [CrossRef]
13. Rong, D.; Wang, H.; Ying, Y.; Zhang, Z.; Zhang, Y. Peach variety detection using VIS-NIR spectroscopy and deep learning. *Comput. Electron. Agric.* **2020**, *175*, 105553. [CrossRef]

14. Fan, S.; Li, J.; Xia, Y.; Tian, X.; Guo, Z.; Huang, W. Long-term evaluation of soluble solids content of apples with biological variability by using near-infrared spectroscopy and calibration transfer method. *Postharvest Biol. Technol.* **2019**, *151*, 79–87. [CrossRef]
15. Chen, H.; Pan, T.; Chen, J.; Lu, Q. Waveband selection for NIR spectroscopy analysis of soil organic matter based on SG smoothing and MWPLS methods. *Chemom. Intell. Lab. Syst.* **2011**, *107*, 139–146. [CrossRef]
16. Hao, Y.; Wu, W.; Shang, Q.; Geng, P. Analysis Model of Oleic and Linoleic Acids in Camellia Oil via Near-Infrared Spectroscopy. *Acta Optica Sinica.* **2019**, *39*, 381–386. [CrossRef]
17. Diwu, P.; Bian, X.; Wang, Z.; Liu, W. Study on the Selection of Spectral Preprocessing Methods. *Spectrosc. Spectral Anal.* **2019**, *39*, 2800–2806. Available online: http://www.gpxygpfx.com/CN/10.3964/j.issn.1000-0593(2019)09-2800-07 (accessed on 27 June 2022).
18. Lao, C.; Chen, J.; Zhang, Z.; Chen, Y.; Ma, Y.; Chen, H.; Gu, X.; Ning, J.; Jin, J.; Li, X. Predicting the contents of soil salt and major water-soluble ions with fractional-order derivative spectral indices and variable selection. *Comput. Electron. Agric.* **2021**, *182*, 106031. [CrossRef]
19. Mansuri, S.M.; Chakraborty, S.K.; Mahanti, N.K.; Pandiselvam, R. Effect of germ orientation during Vis-NIR hyperspectral imaging for the detection of fungal contamination in maize kernel using PLS-DA, ANN and 1D-CNN modelling. *Food Control.* **2022**, *139*, 109077. [CrossRef]
20. de Santana, F.B.; Otani, S.K.; de Souza, A.M.; Poppi, R.J. Comparison of PLS and SVM models for soil organic matter and particle size using vis-NIR spectral libraries. *Geoderma Reg.* **2021**, *27*, e00436. [CrossRef]
21. Yan, K.; Zhou, X. Chiller faults detection and diagnosis with sensor network and adaptive 1D CNN. *Digit. Commun. Netw.* **2022**, *8*, 531–539. [CrossRef]
22. Tian, S.; Wang, S.; Xu, H. Early detection of freezing damage in oranges by online Vis/NIR transmission coupled with diameter correction method and deep 1D-CNN. *Comput. Electron. Agric.* **2022**, *193*, 106638. [CrossRef]
23. Zou, X.; Zhao, J.; Povey, M.J.W.; Holmes, M.; Hanpin, M. Variables selection methods in near-infrared spectroscopy. *Anal. Chim. Acta.* **2010**, *667*, 14–32. [CrossRef]
24. Jaillais, B.; Pinto, R.; Barros, A.S.; Rutledge, D.N. Outer-product analysis (OPA) using PCA to study the influence of temperature on NIR spectra of water. *Vib. Spectrosc.* **2005**, *39*, 50–58. [CrossRef]
25. Jiang, J.; Guo, M.; Yang, S. Fault diagnosis of rolling bearings based on GAF and DenseNet. *J. Mine Autom.* **2021**, *47*, 84–89. [CrossRef]
26. Yao, L.; Sun, J.; Ma, C. Fault Diagnosis Method for Rolling Bearings Based on Grame Angle Fields and CNN-RNN. *Bearing* **2022**, *2*, 61–67. [CrossRef]
27. Liu, S.; Wang, S.; Hu, C.; Bi, W. Determination of alcohols-diesel oil by near infrared spectroscopy based on gramian angular field image coding and deep learning. *Fuel* **2022**, *309*, 122121. [CrossRef]

biosensors

MDPI

Article

A Competitive "On-Off-Enhanced On" AIE Fluorescence Switch for Detecting Biothiols Based on Hg^{2+} Ions and Gold Nanoclusters

Shuqi Li [1,2,†], Yuqi Wan [1,2,†], Yu Li [1,2], Jinghan Liu [1,2], Fuwei Pi [1,2,*] and Ling Liu [3,*]

1 State Key Laboratory of Food Science and Technology, School of Food Science and Technology, Jiangnan University, Wuxi 214122, China
2 Collaborative Innovation Center of Food Safety and Quality Control in Jiangsu Province, Jiangnan University, Wuxi 214122, China
3 Wuxi Institute of Technology, Wuxi 214122, China
* Correspondence: pifuwei@jiangnan.edu.cn (F.P.); liulin@wxit.edu.cn (L.L.)
† These authors contributed equally to this work.

Abstract: In this study, a novel "on-off-enhanced on" approach to highly sensitive rapid sensing of biothiols was developed, based on competitive modulation of gold nanoclusters (AuNCs) and Hg^{2+} ions. In our approach, the AuNCs were encapsulated into a zeolite imidazole framework (ZIF) for predesigned competitive aggregation-induced luminescence (AIE) emission. To readily operate this approach, the Hg^{2+} ions were selected as mediators to quench the fluorescence of AuNCs. Then, due to the stronger affinities between the interactions of Hg^{2+} ions with -SH groups in comparison to the AuNCs with -SH groups, the quenched probe of AuNCs@ZIF-8/Hg^{2+} displayed enhanced fluorescence after the Hg^{2+} ions were competitively interacted with -SH groups. Based on enhanced fluorescence, the probe for AuNCs@ZIF-8/Hg^{2+} had a sensitive and specific response to trace amounts of biothiols. The developed fluorescence strategy had limit of quantification (LOQ) values of 1.0 μM and 1.5 μM for Cys and GSH molecules in serum, respectively. This competitive AIE strategy provided a new direction for developing biological probes and a promising method for quantifying trace amounts of biothiols in serum. It could promote progress in disease diagnosis.

Keywords: aggregation-induced emission; fluorescence enhancement; detection; biothiols; sensitive

check for updates

Citation: Li, S.; Wan, Y.; Li, Y.; Liu, J.; Pi, F.; Liu, L. A Competitive "On-Off-Enhanced On" AIE Fluorescence Switch for Detecting Biothiols Based on Hg^{2+} Ions and Gold Nanoclusters. *Biosensors* **2023**, *13*, 35. https://doi.org/10.3390/bios13010035

Received: 23 November 2022
Revised: 20 December 2022
Accepted: 24 December 2022
Published: 27 December 2022

1. Introduction

Biothiols, as essential signaling molecules, e.g., cysteine (Cys) and glutathione (GSH), play important roles in physiological activities [1]. Several studies have confirmed that abnormal levels of biothiols were highly associated with neurotoxicity and cardiovascular disease [2–4]. For example, a high level of homocysteine is a reliable risk factor for hypertension, stroke and heart attack [5,6]. Significantly reduced GSH levels in serum can be used as diagnostic signs of early Parkinson's disease. Levels of GSH are related to neurodegenerative diseases, diabetes, HIV infection and cancer [7–9]. In addition to the above diseases, sudden changes in Cys molecule levels are potentially associated with liver damage, skin diseases and Alzheimer's disease [10–12]. Therefore, the necessary point-of-care-testing (POCT) techniques relating to evaluation of biothiol levels—especially visual approaches—have attracted more attention in analytical sciences.

Currently, traditional methods, such as capillary electrophoresis [13], high performance liquid chromatography [14], colorimetry [15] and mass spectrometry [16] are widely used in the evaluation of biothiols. However, expensive equipment, time-consuming operation and complex preparations highly limit onsite applications. Over the past decade, rapid fluorescence detection technology, based on various fluorescent probes, has developed rapidly, providing simple, robust, and fast sensing strategied for the quantitative analysis and monitoring of biological substances, biologically relevant ions and other trace substances [17–19]. Nanoprobes, as another alternative approach, have also been seen as promising strategies for rapid sensing of biothiols [20,21]. Among these, gold nanoclusters (AuNCs) have been marked as significant probes on fluorescence output due to their simple synthetic methods, good photostability, rapid sensing, chemical stability and biocompatibility [22]. Notwithstanding, to date, most synthetic techniques for AuNCs have displayed lower fluorescence quantum yields compared to common fluorescent dyes such as rhodamine [23,24]. Thus, a promising approach for modulating the fluor-features of AuNCs has become necessary and desirable for Au-based POCT techniques.

As a class of porous crystalline materials, zeolite imidazole frameworks (ZIF-MOFs) have been used to adsorb and separate various ions or applied as a carrier to co-encapsulate molecular probes, such as enzymes, carbon dots and gold nanomaterials [25–28]. Due to the confining holes of ZIF-MOFs, the encapsulated molecules usually generate more effective performances than those of which they are capable. For example, Jingtian Chi et al. [29] incorporated AuNCs with glucose oxidase (GO) into ZIF-8, which greatly improved the catalytic activity of the enzymes and the storage stability of the nanomaterials compared to the free GOx/AuNCs solution under the protection of ZIF-8 scaffold. For highly effective detection of tetracycline residues in milk, Alireza Khataee et al. [30] encapsulated both AuNCs and copper nanoclusters (CuNCs) into ZIF-8. Based on their construction, the quantum yields of BSA-AuNCs were improved fifteenfold due to restricted molecular motion and a 4.8 nM of the limit of detection was obtained. The significantly increased emission efficiency of AuNCs and the protective effects of the ZIF-MOFs shell couldpave a promising way forward for further application of AuNCs in hazards, specifically in molecule sensing and sensitive POCT evaluations.

In the present study, a novel "on-off-enhanced on" approach was developed, based on the competitive modulation of AuNCs in ZIF-8-MOF for highly sensitive rapid sensing of biothiols. As illustrated in Scheme 1, in our strategy, the AuNCs were encapsulated into ZIF-8-MOF. Due to the self-aggregation of AuNCs, an intensive aggregation-induced luminescence was achieved. Differing from conventional AIE ideas, the aggregated luminescence was modulated and quenched through a pre-burying competitive factor of Hg^{2+} ions to develop a visual-sensing nanoprobe. In the presence of biothiols, the mediators of Hg^{2+} ions were competed from AuNCs, based on the formation of stronger Hg^{2+}-S bonds, which resulted in the recovery and generation of enhanced fluorescence. A nanoprobe such as that devised could successfully respond to trace levels of biothiols in serum samples. Notably, the solid stability of free AuNCs also demonstrated great potential for sensing biothiols and could promote progress in disease diagnosis.

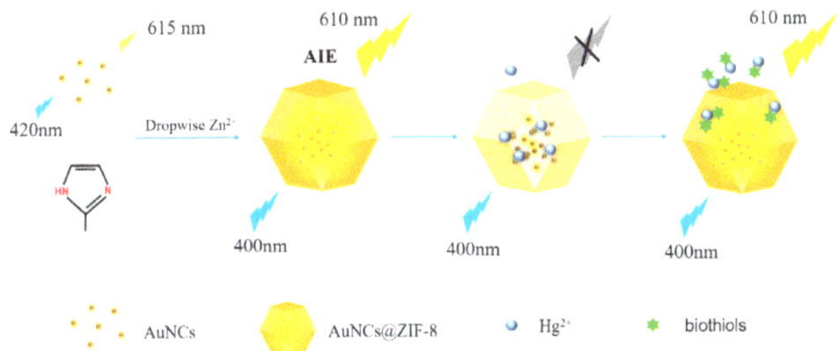

Scheme 1. Schematic illustration of in situ self-assembly of AuNCs@ZIF-8 and the sensing mechanism of biothiols by AuNCs@ZIF-8 and Hg^{2+} ions.

2. Experimental Section

2.1. Materials and Apparatus

Chloroauric acid trihydrate ($HAuCl_4{\cdot}3H_2O$, purity > 99.9%) was supplied by Adamas Reagent Co., Ltd. (Shanghai, China). Glutathione reduced (GSH) was purchased from Innochem Co., Ltd. (Beijing, China). 2-Methylimidazole (2-MIM) was purchased from J&K Scientific Co., Ltd. (Shanghai, China). Zinc nitrate hexahydrate ($Zn(NO_3)_2$, purity > 99%), L-Cysteine(Cys), L-Aspartic acid (Asp), L-Arginine (Arg), L-Lysine (Lys), Glycine (Gly),

L-Valine (Val), L-Isoleucine (Ile), L-Leucine (Leu), L-Histidine (His), L-Phenylalanine (Phe), L-Threonine (Thr), L-Proline (Pro), L-Alanine (Ala), L-Glutamic acid (Glu), L-Tryptophan (Trp), L-Serine(Ser), L-Methionine (Met), L-Glutamine (Gln) L-Asparagine (Asn) and tyrosine (Tyr) were purchased from Sinopharm Chemical Reagent Co., Ltd. (Shanghai, China).

All chemical reagents were AR-grade and used without further purification. The ultrapure water used throughout the experiments was obtained from a Millipore water purification system (18.2 MΩ, Milli-Q, Millipore, Burlington, MA, USA).

2.2. Synthesis of AuNCs

AuNCs were synthesized according to a previously reported method with minor modification [31]. Briefly, an aqueous solution of $HAuCl_4 \cdot 3H_2O$ (20 mM, 0.5 mL) was mixed with 4.35 mL ultrapure water. Then, GSH solution (100 mM, 0.15 mL) was slowly dripped into $HAuCl_4 \cdot 3H_2O$ solutions at 25 °C. After 5 min of magnetic stirring at 500 rpm, the temperature was increased to 70 °C and stirring continued for 24 h. Finally, the AuNCs solution was concentrated to 1.5 times and stored at 4 °C for the following experiments. All glass bottles were washed and soaked in aqua regia before use.

2.3. Synthesis of AuNCs@ZIF-8 Nanocomposites

One-pot synthesis of the AuNCs@ZIF-8 nanocomposites was prepared by mixing concentrated AuNCs solution with the precursors of ZIF-8. That is, the $Zn(NO_3)_2 \cdot 6H_2O$ (0.08 M, 0.5 mL) and 2-MIM (4 M, 3.5 mL) was first dissolved in the concentrated solution of AuNCs solution to obtain solutions A and B, respectively. Then, solution A was dripped into solution B under magnetic stirring (500 rpm) at 25 °C. After 30 min reaction, nanocomposites were sat for 2 h and washed three times under centrifugation (10,000 rpm, 10 min).

2.4. Characterization of AuNCs and AuNCs@ZIF-8

Transmission electron microscopy (TEM) images were acquired through a JEOL JEM-2100 transmission electron microscope with an acceleration voltage of 200 kV. UV-vis spectra for AuNCs, ZIF-8 and AuNCs@ZIF-8 in the range of 800–250 nm with scan interval 1 nm were obtained on a T9 UV-vis spectrophotometer provided by Beijing Purkinje General Instrument Co. Beijing China. The crystal structure of ZIF-8 and AuNCs@ZIF-8 were obtained on a D2 PHASER X-ray diffractometer (XRD; Karlsruhe, Germany) using Cu-$K\alpha$ radiation as the X-ray source at 30 kV and 10 mA. Fourier transform infrared spectra (FT-IR) in the range of 400–4000 cm^{-1} was measured on an IS10 IR spectrophotometer (Nicolet, Waltham, MA, USA) using KBr pellet method. The fluorescence spectra were observed on a Fluoro Max4 (Horiba JY, Irvine, CA, USA).

To evaluate the fluorescence enhancement effects of AuNCs@ZIF-8, the fluorescence quantum yields of freshly-synthesized AuNCs solution and AuNCs@ZIF-8 synthesized with 1.5 times concentrated AuNCs were measured. Differences in fluorescence intensity of AuNCs and AuNCs@ZIF-8 were additionally measured after dilution to equal AuNCs concentrations. To evaluate the temporal and Ph stability of the prepared AuNCs and AuNCs@ZIF-8, AuNCs and AuNCs@ZIF-8, stored at 4 °C, were taken out every 1 month to measure fluorescence intensity. Changes in fluorescence intensity with Ph of the solutions were measured by respectively taking and diluting 20 μL of AuNCs and AuNCs@ZIF-8 to 1 mL with different Ph of water (pH = 4.5, 5, 6, 7, 8). To investigate the "on-off-enhanced on" fluorescence switch, the fluorescence intensity values of 5 μL of AuNCs@ZIF-8 before and after quenching by Hg^{2+}, were recorded, and the fluorescence intensity values were recorded with the addition 1 mL of Cys solution.

2.5. Optimization of Reaction Conditions

To determine the optimal amount of Hg^{2+}, 10 μL of AuNCs@ZIF-8 was added to different volumes of Hg^{2+} solution (3×10^{-4} M, 0 μL, 5 μL, 10 μL, 12 μL, 15 μL, 16 μL, 18 μL, 20 μL and 30 μL) and the fluorescence intensity was measured after diluting to 1 mL system with water. To determine the reaction time of AuNCs@ZIF-8 with Hg^{2+}, 10 μL of

AuNCs@ZIF-8 was added to the optimal amount of Hg^{2+}, and the change of fluorescence intensity with time was measured after diluting to 1 mL system with water.

To evaluate the performance of as-prepared nanoprobes, the volume ratio of AuNCs@ZIF-8 to Hg^{2+} ions (3×10^{-4} M) was first fixed. Then, the volume of AuNCs@ZIF-8 was changed and the solution systems were replenished to the same volume with ultrapure water. After adding Cys solution (5 μM, 1 mL) and reacting for 15 min, the fluorescence spectrum of the solution was measured under 400 nm excitation. The degree of fluorescence recovery of AuNCs@ZIF-8/Hg^{2+} after the addition of biothiol concentration was expressed by Equation (1):

$$\frac{(FL - FL_0)}{FL_0} \tag{1}$$

Here, FL_0 and FL are the fluorescence intensities of the AuNCs@ZIF-8/Hg^{2+} in the absence and presence of biothiols, respectively.

The optimal reaction time of Hg^{2+}/AuNCs@ZIF-8 with Cys was evaluated by adding Cys (1 mL, 5 μM) to Hg^{2+}/AuNCs@ZIF-8 and recording the change of fluorescence intensity with time.

2.6. Fluorescent Detection of Cys and GSH Molecules

Under shaking, 1 mL of Cys solution at different concentrations (0.0, 1.0, 1.5, 2.0, 2.5, 4.0, 5.0, 8.0, 10.0, 20.0 and 30.0 μM) was separately added into a 5 μL of AuNCs@ZIF-8 solution containing Hg^{2+} ions (3×10^{-4} M, 8 μL). Finally, the fluorescence spectra were measured by a fluorescence spectrometer. Parameter settings: excitation wavelength of 500–750 nm, scanning speed at 1 nm/min. Same parameters were used to detect GSH. Moreover, the detection conditions for Cys molecules were optimized accordingly, mainly including dosage of AuNCs@ZIF-8 probe, the dosage ratio of Hg^{2+} ions to AuNCs@ZIF-8 and the incubation time of Cys and Hg^{2+} ions. In addition, the control tests were separately conducted, accordingly, for different analytes (10 folds the Cys solution) of other amino acids (Asp, Arg, Lys, Gly, Val, Ile, Leu, His, Phe, Thr, Pro, Ala, Glu, Trp, Ser, Met, Gln, Asn and Tyr) and GSH. Detection of Cys molecules with AuNCs/Hg^{2+} followed the same detection conditions as AuNCs@ZIF-8/Hg^{2+}.

2.7. Real Sample Analysis

For the real sample test, bovine serum was first treated with acetonitrile at a volume ratio of 1:5, shaking for 30 s and leaving it to rest for 15 min. After centrifugation to remove protein precipitation at 10,000 rpm for 20 min, acetonitrile was removed by nitrogen blowing and diluted to 200 folds with ultrapure water as solvent. Different concentrations of Cys molecules were pre-prepared with the treated serum for spiked recovery experiments.

3. Result and Discussion

3.1. Characterization of GSH-AuNCs and AuNCs@ZIF-8

The structure of GSH-AuNCs and AuNCs@ZIF-8 particles were characterized using TEM (Figure 1). As shown in Figure 1A, the original product of GSH-AuNCs particles was a spherical shape with good dimensional uniformity. The mean particle size was 2.21 ± 0.49 nm, calculated from 150 randomly selected particles (Figure 1A insert). Emission fluorescence of GSH-AuNCs showed an optimum at 615 nm under the excitation of 420 nm (Figure S1A in Supplementary Materials). Moreover, when the AuNCs solution was irradiated by a UV light at 365 nm, the color of solution changed from a pale yellow to a luminous orange-yellow (Figure S1A insert). However, compared with ZIF-8 MOFs, which have a regular uniform polyhedral shape with a uniform size distribution of ca. 120 nm (Figure 1B insert), the one-pot synthesized AuNCs@ZIF-8 particles only showed slight increases in shape and size. Moreover, the zoom-in images, as plotted in Figure 1C, clearly showed that the AuNCs particles were encapsulated inside of the ZIF-8 framework rather than adsorbed on the surface. Meanwhile, UV-vis spectra of AuNCs, ZIF-8 and AuNCs@ZIF-8 particles also indicated that the absorption of AuNCs in AuNCs@ZIF-8 particles was not affected, i.e., the ZIF-8 MOFs had no absorption peak in the range of 250–800 nm, while both AuNCs and AuNCs@ZIF-8 had clear absorption peaks at 400 nm (Figure S1B).

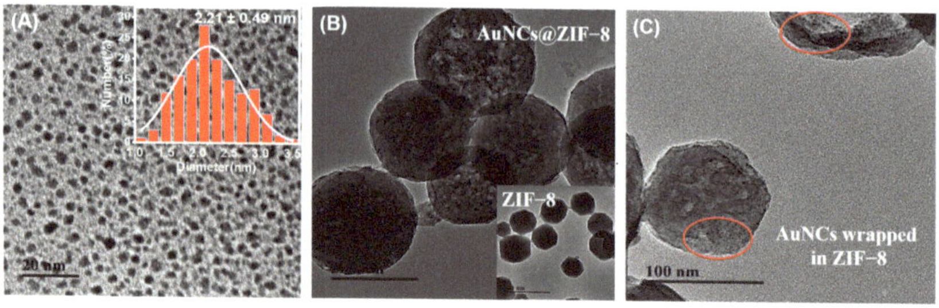

Figure 1. TEM morphology of (**A**) as-prepared GSH-AuNCs(insert size distribution of as-prepared GSH-AuNCs), (**B**) free ZIF-8 and AuNCs@ZIF-8 nanoparticles and (**C**) the AuNCs wrapped in ZIF-8 which were marked in red circles.

The fidelity of the ZIF-8 crystalline structure after imbedding of AuNCs were confirmed with XRD (Figure 2A). The XRD patterns of ZIF-8 and AuNCs@ZIF-8 exhibited characteristic and sharp peaks of pure-phase ZIF-8. It was observed that theAuNCs@ZIF-8 could exhibit no significant change in the phase structures. As plotted in Fourier transform infrared spectroscopy (FT-IR) (Figure 2B), the peaks at 2500–3600 cm^{-1} and 1640–1660 cm^{-1} corresponded to stretching vibrations of -OH groups and C=O stretching in AuNCs, respectively. The characteristic absorption bands of -OH disappeared from the FT-IR spectrum of the AuNCs@ZIF-8 particle. Such obvious variances implied the coordination of -COOH group in GSH with Zn^{2+} ion existing in the ZIF-8 framework. On the other side, an additional characteristic absorption band of AuNCs@ZIF-8 particle at 1640–1660 cm^{-1} also indicated the presence of AuNCs in ZIF-8.

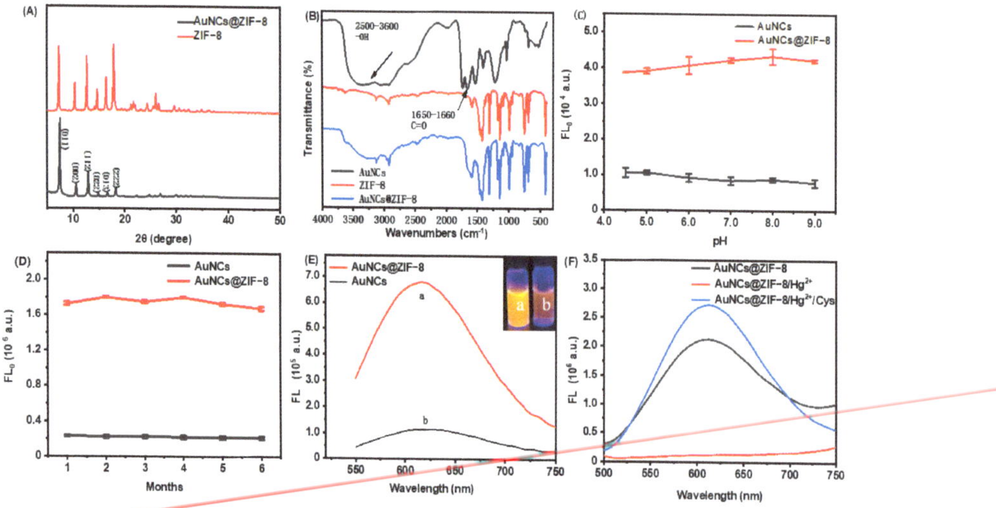

Figure 2. Basic properties of ZIF-8 and AuNCs@ZIF-8 MOFs. (**A**) XRD pattern, (**B**) FT-IR spectra, effect of pH (**C**) and time (**D**) on material stability, (**E**) fluorescence emission spectra containing same concentration of AuNCs. (**F**) The fluorescence emission spectra of AuNCs@ZIF-8, AuNCs@ZIF-8/Hg^{2+} and AuNCs@ZIF-8/Hg^{2+}/Cys.

We investigated the fluorescence stability of AuNCs and AuNCs@ZIF-8 with pH and storage time (Figure 2C,D). We found that AuNCs, before and after coating by ZIF-8, did not show significant change in the pH range of 4.5–9. Thus, they presumably had good Ph stability in serum and other detection systems. In addition, the fluorescence intensity of AuNCs and AuNCs@ZIF-8 was not significantly lost during storage of 4 °C for six months.

In addition to the few changes occurring between ZIF-8 and AuNCs@ZIF-8 particles, interestingly, the fluorescence intensity of AuNCs@ZIF-8 (a) was substantially enhanced (approximately sixfold) in comparison to the AuNCs solution (b) with same molarity (Figure 2E insert). It was demonstrated that the enhanced luminescence of AuNCs was linked to the reduced nonradiative decay caused by restriction of intramolecular movements [32]. Thus, it was highly reasonable to speculate that the limited space in ZIF-8 MOFs blocked the intramolecular rotation and provided an unfirm aggregation of AuNCs, which finally resulted in the phenomenon of aggregation-induced emission (AIE) [33–35]. The enhanced features were also proven by fluorescence quantum yields and were consistent with previous reports [36]. That is, after optimization of the synthesis conditions of AuNCs@ZIF-8 particles, the quantum yields of primary AuNCs solution and synthesized AuNCs@ZIF-8 particles were 0.75% and 20.65%, respectively. To effectively sense hazards, synthesized AuNCs were concentrated and used at optimum concentration, i.e., 1.5 times in 2-MIM solution (Figure S2).

As plotted in Figure 2F, although the fluorescence intensity of AuNCs@ZIF-8 was intensively improved, it rapidly decreased when Hg^{2+} ions were added. This was probably due to the formation of a metalophulid bond between Hg^{2+} ions and Au^{+} existing on the surface of AuNCs [31,37]. On the other hand, the quenched emissions gradually recovered when the particles met with Cys molecules. Such lost-and-found fluorescence probably originated from the higher, more stable reactions between Hg^{2+} ions and sulfhydryl, compared with the reactions between AuNCs and Hg^{2+} ions [38].

3.2. Optimization of Experimental Conditions

To handily control and operate the as-prepared "on-off-enhanced on" approach, the effects of synthesis conditions, including the synthesis time, the volume of GSH solution and the synthesis temperature of fluorescence performances of AuNCs were investigated and the results are plotted in Figure S3. As demonstrated, the fluorescence intensity of AuNCs gradually increased as the reaction kept going, and the optimal intensity was obtained after 24 h (Figure S3A). However, when the temperature exceeded 70 °C, the reaction gradually suspended, even though the maximum GSH was set optimally at 150 μL (Figure S3B).

In addition to the synthesis conditions, assay operation factors, such as the amount of fluorescence quenching effect observed with Hg^{2+}, AuNCs@ZIF-8 composite and the incubation time of biothiols, were also explored to thoroughly elucidate the sensing principles. As shown in Figure 3A, it was observed that the fluorescence intensity gradually decreased with the increase of Hg^{2+} dosage. Such variance implied that more Hg^{2+} could induce a more stable structure of Hg^{2+}/Au^{+}, but excessive amounts of Hg^{2+} could finally react with biothiols, resulting in a lack of fluorescence response to the target. To obtain optimal detection sensitivity and a wide detection range, the amount of Hg^{2+} was set at 16.0 μL (approximately 90% quenching of initial fluorescence). Under the optimal volume of Hg^{2+}, as-prepared probes also displayed an AuNCs@ZIF-8 responding, as illustrated in Figure 3B. With trace numbers of probes, it was difficult to provide good fluorescence; however, an excess of probes undoubtedly reduced the sensitivity of detection.

Figure 3C,D plotted the time-dependent fluorescence quenching of AuNCs@ZIF-8 to Hg^{2+} ions and fluorescence restoration of the probe of AuNCs@ZIF-8/Hg^{2+} to Cys molecules, respectively. The fluorescence intensity of AuNCs@ZIF-8 particles was significantly quenched in 1 min after meeting with Hg^{2+} ions, and the fluorescence intensity reached a stable level after 5 min. When Cys molecules were present in the solution, the fluorescence intensity of AuNCs@ZIF-8/Hg^{2+} probes sharply increased, reaching the maximum after 10 min. Such a rapid response of our AuNCs@ZIF-8/Hg^{2+} probes demonstrated great potential for onsite POCT detection of biothiols.

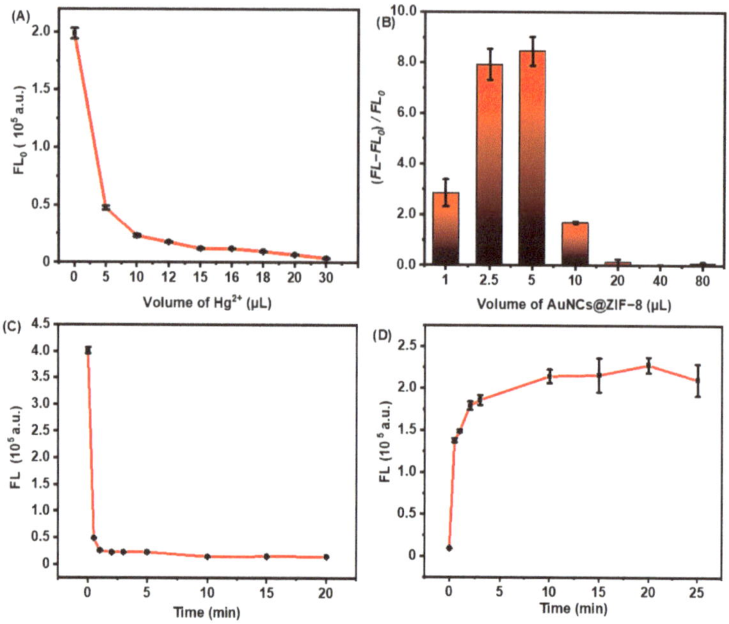

Figure 3. Fluorescence changes of as-prepared sensing probes. (**A**) The fluorescence intensity of AuNCs@ZIF-8 at 400 nm wavelength with different concentrations of Hg^{2+}. (**B**) Fluorescence changes of the same concentration of Cys detected by different volumes of AuNCs@ZIF-8. (**C**) Relationship between the fluorescence intensity of AuNCs@ZIF-8 and time for adding Hg^{2+} ions. (**D**) Relationship between the fluorescence intensity of AuNCs@ZIF-8/Hg^{2+} and time for adding Cys.

3.3. Sensitivity and Selectivity of AuNCs@ZIF-8/Hg^{2+} Probes on Cys and GSH Molecules

To investigate the specific features of as-prepared AuNCs@ZIF-8/Hg^{2+} probes on biothiol sensing, the responses of GSH and human amino acids, including Cys molecules, were tested. As plotted in Figure 4, not all of the amino acids displayed evident fluorescence recovery. As is known, Cys molecules are the only amino acid containing -SH groups among the 20 amino acids necessary for humans, while GSH is a thiol-containing tripeptide composed of Cys molecules, glutamic acid and glycine. Therefore, the AuNCs@ZIF-8/Hg^{2+} probes also displayed restored fluorescence on GSH molecules based on interactions between -SH groups and Hg^{2+} ions. However, probably due to the spatial block of tripeptide, the fluorescence intensity of GSH molecules was lower than that of free Cys molecules. The excellent specificity of AuNCs@ZIF-8/Hg^{2+} probes on detection of Cys and GSH molecules suggested that our developed AuNCs@ZIF-8/Hg^{2+} approach exhibited great potential to sense Cys and GSH molecules.

3.4. Quantitative Sensing on Trace Amount of Cys and GSH Molecules

The quantitative evaluation of Cys and GSH molecules with our developed AuNCs@ZIF-8/Hg^{2+} approach was studied by measuring the fluorescence intensity at 400 nm. As plotted in Figure 5, after Hg^{2+} was mixed with AuNCs@ZIF-8, it was added to different concentrations of Cys or GHS solutions and incubated for 10 min for fluorescence testing. Cys and GHS molecules were able to establish a series of good linear relationships with AuNCs@ZIF-8-based probes. The detection of Cys molecules with AuNCs@ZIF-8 probes had a linear relationship in the range of 1–10 μM, and the linear equation was y = 2.083x − 2.907 with R^2 = 0.991.

Similarly, the detection of GSH with AuNCs@ZIF-8 probes had a linear relationship in the range of 1.5–8 μM, and the linear equation was y = 0.950x − 1.23 with R^2 =0.999. Furthermore, compared with previous reports on Cys and GSH molecules, as summarized in Table 1, our AuNCs@ZIF-8/Hg^{2+} strategy displayed a much better response tos trace amount of Cys and GSH molecules regarding linear ranges and LODs, simultaneously.

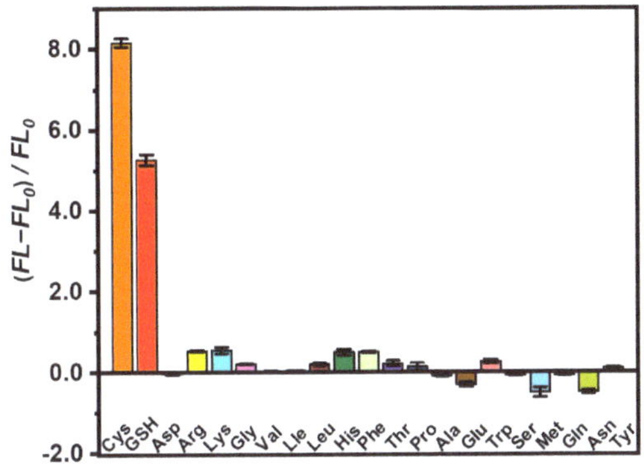

Figure 4. Fluorescence intensity change of AuNCs@ZIF-8 at λ_{ex} = 400 nm in the presence of various amino acids.

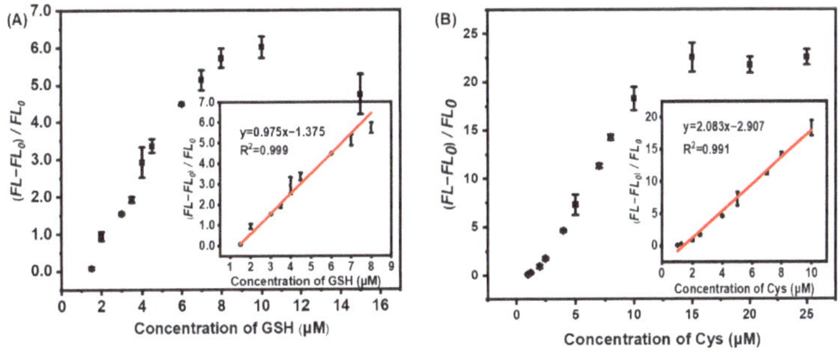

Figure 5. Relationship of fluorescence change of AuNCs@ZIF-8 and concentration of (**A**) GSH and (**B**) Cys molecules.

Table 1. Comparison of different fluorescent probes for detection of biothiols.

Material	Linear Range	LOD	Reference
CDs	2–20 μM for Cys	0.29 μM	[39]
AuNCs/AuNPs	1.5–35 μM for Cys	1.4 μM	[40]
AuNCs	8.3–100 μM for Cys	1.45 μM	[41]
AuNCs	0.2–60 μM for Cys	80 nM	[42]
AgNCs/NCDs	20–80 μM for Cys 20–80 μM for GSH	0.14 μM 0.4 μM	[43]
CeO_2/CoO	5–10 μM for Cys	3.71 μM	[44]
AuNCs@ZIF-8	1–10 μM for Cys 1.5–8 μM for GSH	0.15 μM 0.32 μM	Our work

3.5. Detection Performance in Real Samples

To further investigate the feasibility of using AuNCs@ZIF-8/Hg^{2+} probes to sense biothiols in blood sera, recovery experiments ON biothiols in serum samples were carried out. Low, medium and high concentrations of biothiols were selected to verify the universality of our strategy. For the sample preparation, after the protein from bovine serum was precipitated, the supernatant was diluted to a certain concentration, as standard for adding of Cys or GSH molecules, respectively. The results of quantitative evaluation are depicted in Table 2. As listed in Table 2, the recovery of spiked Cys molecules ranged from 97.36% to 115.31%, while that of spiked GSH molecules ranged from 105.29% to 115.46% with an RSD less than 5%. Such quantitative results implied that our developed AuNCs@ZIF-8/Hg^{2+} strategy provided a promising approach to accurate and rapid evaluation of biothiols in blood serum.

Table 2. The recovery of Cys and GSH molecules detection in real samples.

Types of Biothiols	Detected (μM)	Spiked (μM)	Found (μM)	Recovery (%)	RSD (n = 3, %)
		2	3.25 ± 0.06	111.94 ± 2.86	2.56
Cys	1.008	4	4.90 ± 0.15	97.36 ± 3.83	3.94
		8	10.39 ± 0.12	115.31 ± 3.37	2.92
		2	3.64 ± 0.14	115.46 ± 4.24	3.67
GSH	1.567	4	6.09 ± 0.01	113.04 ± 0.36	0.32
		6	7.88 ± 0.20	105.29 ± 3.40	3.22

4. Conclusions

In summary, a "on-off-enhanced on" strategy for sensitive and fast sensing of biothiols in blood serum was developed based on AuNCs@ZIF-8 fluorescence nanoparticles. By combining the ZIF-8 framework with mediators of Hg^{2+} ions, our approach displayed AIE behavior for AuNCs under the embedding of ZIF-8 framework. However, the fluorescence intensity sharply decreased when meeting with mediators of Hg^{2+} ions. Due to the different affinities between the interactions of Hg^{2+} ions with -SH groups and the AuNCs with -SH groups, the quenched probe of AuNCs@ZIF-8/ Hg^{2+} displayed an obviously enhanced fluorescence after the Hg^{2+} ions were competitively interacted with -SH groups. Differing from conventional AIE strategy, our competitively-enhanced AIE behavior resulted in a much lower LOD and longer fluorescence lifetime on sensing trace amounts of biothiols. In light of these findings, this facile, accurate, and reliable competitive AIE strategy could provide a promising alternative method of quantifying trace amounts of biothiols in sera and biological systems.

Supplementary Materials: The following supporting information can be downloaded at: https://www.mdpi.com/article/10.3390/bios13010035/s1, Figure S1. (A) Fluorescence excitation and emission spectra of as-prepared GSH-AuNCs. (B) UV-vis spectra of ZIF-8, AuNCs and AuNCs/ZIF-8; Figure S2. The fluorescence emission spectra of AuNCs/ZIF-8 synthesized from (A) different concentration of 2-MIM solution and (B) different concentration times of AuNCs solution. The red broken line represents the coverage of AuNCs.; Figure S3. The effect of (A) synthesis time (B) Volume of GSH solution and (C) synthetic temperature on the fluorescence intensity of AuNCs.

Author Contributions: S.L. and Y.W.: investigation lead, writing—original draft; Y.L.: investigation, methodology; J.L.: writing—review and editing. F.P.: conceptualization lead, funding acquisition, project administration, supervision lead, writing—original draft and review. L.L.: conceptualization lead, investigation, supervision lead, and review. All authors have read and agreed to the published version of the manuscript.

Funding: This project was supported by the Innovation and Entrepreneurship Projects and Six Talent Peaks Project of Jiangsu Province (No. SWYY-023); F.P. thanks for the support from the National Natural Science Foundation of China (No. 21603087) and High-level Overseas Talent Workstation of Shandong Province.

Institutional Review Board Statement: Not applicable.

Informed Consent Statement: Not applicable.

Data Availability Statement: The data presented in this study are available on request from the corresponding author.

Conflicts of Interest: There authors declare no conflict of interest.

References

1. Ulrich, K.; Jakob, U. The role of thiols in antioxidant systems. *Free Radic. Biol. Med.* **2019**, *140*, 14–27. [CrossRef] [PubMed]
2. Townsend, D.M.; Tew, K.D.; Tapiero, H. The importance of glutathione in human disease. *Biomed. Pharmacother.* **2003**, *57*, 145–155. [CrossRef] [PubMed]
3. Boldyrev, A. Molecular mechanisms of homocysteine toxicity and possible protection against hyperhomocysteinemia. *Free Radic. Biol. Med.* **2009**, *47*, 138–139.
4. Kalra, D.K. Homocysteine and cardiovascular disease. *Clin. Appl. Immunol. Rev.* **2004**, *6*, 101–106. [CrossRef]
5. Baszczuk, A.; Kopczyński, Z. Hyperhomocysteinemia in patients with cardiovascular disease. *Postep. Ostep. Hig. Med. Dosw.* **2014**, *68*, 579–589. [CrossRef]
6. Perry, I.J. Homocysteine, hypertension and stroke. *J. Hum. Hypertens.* **1999**, *13*, 289–293. [CrossRef]
7. Morris, D.; Ly, J.; Chi, P.T.; Daliva, J.; Nguyen, T.; Soofer, C.; Chen, Y.C.; Lagman, M.; Venketaraman, V. Glutathione synthesis is compromised in erythrocytes from individuals with HIV. *Front Pharmacol.* **2014**, *5*, 73. [CrossRef]
8. Kim, A.D.; Zhang, R.; Han, X.; Kang, K.A.; Piao, M.J.; Maeng, Y.H.; Chang, W.Y.; Hyun, J.W. Involvement of glutathione and glutathione metabolizing enzymes in human colorectal cancer cell lines and tissues. *Mol. Med. Rep.* **2015**, *12*, 4314–4319. [CrossRef]
9. Feng, G.; Chauhan, V.; Chauhan, A. Glutathione redox imbalance in brain disorders. *Curr. Opin. Clin. Nutr. Metab. Care* **2014**, *18*, 89–95.
10. Yuan, X.; Tay, Y.; Dou, X.; Luo, Z.; Leong, D.T.; Xie, J. Glutathione-protected silver nanoclusters as cysteine-selective fluorometric and colorimetric probe. *Anal. Chem.* **2013**, *85*, 1913–1919. [CrossRef]
11. Shahrokhian, Saeed, Lead phthalocyanine as a selective carrier for preparation of a cysteine-selective electrode. *Anal. Chem.* **2013**, *85*, 1913–1919.
12. Li, R.; Li, W.; Chen, R.; Lin, W. An activatable water-soluble photoacoustic probe for real-time imaging of endogenous cysteine in the mouse tumor model. *Sens. Actuators B Chem.* **2021**, *347*, 130616. [CrossRef]
13. Isokawa, M.; Kanamori, T.; Funatsu, T.; Tsunoda, M. Analytical methods involving separation techniques for determination of low-molecular-weight biothiols in human plasma and blood. *J. Chromatogr. B Anal. Technol. Biomed. Life Sci.* **2014**, *964*, 103–115. [CrossRef] [PubMed]
14. Manda, K.R.; Adams, C.; Ercal, N. Biologically important thiols in aqueous extracts of spices and evaluation of their in vitro antioxidant properties. *Food Chem.* **2010**, *118*, 589–593. [CrossRef]
15. Xu, H.; Wang, Y.; Huang, X.; Li, Y.; Zhang, H.; Zhong, X. Hg^{2+}-mediated aggregation of gold nanoparticles for colorimetric screening of biothiols. *Analyst* **2012**, *137*, 924–931. [CrossRef]
16. Tsai, C.J.; Liao, F.Y.; Weng, J.R.; Feng, C.H. Tandem derivatization combined with salting-out assisted liquid-liquid microextraction for determination of biothiols in urine by gas chromatography-mass spectrometry. *J. Chromatogr. A* **2017**, *1524*, 29–36. [CrossRef]
17. Pandith, A.; Seo, Y.J. Label-free sensing platform for miRNA-146a based on chromo-fluorogenic pyrophosphate recognition. *J. Inorg. Biochem.* **2020**, *203*, 110867. [CrossRef]
18. Kim, B.Y.; Pandith, A.; Cho, C.S.; Kim, H.S. Highly selective fluorescent probe based on 2-(2′-dansylamidophenyl)-thiazole for sequential sensing of copper(ii) and iodide ions. Bulletin Korean. *Chem. Soc.* **2019**, *40*, 163–168.
19. Pandith, A.; Bhattarai, K.R.; Siddappa, R.K.G.; Chae, H.J.; Seo, Y.J. Novel fluorescent C-2-symmetric sequential on-off-on switch for Cu^{2+} and pyrophosphate and its application in monitoring of endogenous alkaline phosphatase activity. *Sens. Actuator B* **2019**, *282*, 730–742. [CrossRef]
20. Li, H.J. A Review of Off-On Fluorescent nanoprobes: Mechanisms, properties, and applications. *J. Biomed. Nanotechnol.* **2021**, *17*, 1249–1272.
21. Bigdeli, A.; Ghasemi, F.; Abbasi-Moayed, S.; Shahrajabian, M.; Fahimi-Kashani, N.; Jafarinejad, S.; Farahmand Nejad, M.A.; Hormozi-Nezhad, M.R. Ratiometric fluorescent nanoprobes for visual detection: Design principles and recent advances-A review. *Anal. Chim. Acta* **2019**, *1079*, 30–58. [CrossRef] [PubMed]
22. Halawa, M.I.; Lai, J.; Xu, G. Gold nanoclusters: Synthetic strategies and recent advances in fluorescent sensing. *Mater. Today Nano* **2018**, *3*, 9–27. [CrossRef]
23. Chen, L.Y.; Wang, C.W.; Yuan, Z.; Chang, H.T. Fluorescent gold nanoclusters: Recent advances in sensing and imaging. *Anal. Chem.* **2015**, *87*, 216–229. [CrossRef] [PubMed]
24. Tian, R.; Zhang, S.T.; Li, M.W.; Zhou, Y.Q.; Lu, B.; Yan, D.P.; Wei, M.; Evans, D.G.; Duan, X. Localization of Au nanoclusters on layered double hydroxides nanosheets: Nonfinement-induced emission enhancement and temperature-responsive luminescence. *Adv. Funct. Mater.* **2015**, *25*, 5006–5015. [CrossRef]
25. Zhang, X.; Li, Z.; Zhang, T.; Chen, J.; Ji, W.; Wei, Y. Fabrication of an efficient ZIF-8 alginate composite hydrogel material and its application to enhanced copper(ii) adsorption from aqueous solutions. *N. J. Chem.* **2021**, *45*, 15876–15886. [CrossRef]
26. Zhu, G.; Zhang, M.; Lu, L.; Lou, X.; Dong, M.; Zhu, L. Metal-organic framework/enzyme coated optical fibers as waveguide-based biosensors. *Sens. Actuators B Chem.* **2019**, *288*, 12–19. [CrossRef]
27. Tan, Q.; Zhang, R.; Zhang, G.; Liu, X.; Qu, F.; Lu, L. Embedding carbon dots and gold nanoclusters in metal-organic frameworks for ratiometric fluorescence detection of Cu(2). *Anal. Bioanal. Chem.* **2020**, *412*, 1317–1324. [CrossRef]
28. Dai, X.Q.; Hang, Y.H.; Zhao, X.; Sun, C.; Ju, L.N.; Gui, X.J.; Sun, S.; Xu, Y.B.; Wang, Y.R.; Li, Y.F. ZIF-8 as an adsorbent of aqueous phase for Eu and Tb ions. *Micro Nano Lett.* **2016**, *12*, 187–190. [CrossRef]
29. Chi, J.; Guo, M.; Zhang, C.; Zhang, Y.; Ai, S.; Hou, J.; Wu, P.; Li, X. Glucose oxidase and Au nanocluster co-encapsulated metal-organic frameworks as a sensitive colorimetric sensor for glucose based on a cascade reaction. *N. J. Chem.* **2020**, *44*, 13344–13349. [CrossRef]

30. Khataee, A.; Jalili, R.; Dastborhan, M.; Karimi, A.; Azar, A.E.F. Ratiometric visual detection of tetracycline residues in milk by framework-enhanced fluorescence of gold and copper nanoclusters, Spectrochim. *Acta. A Mol. Biomol. Spectrosc.* **2020**, *242*, 118715. [CrossRef]
31. Luo, Z.; Yuan, X.; Yu, Y.; Zhang, Q.; Leong, D.T.; Lee, J.Y.; Xie, J. From aggregation-induced emission of Au(I)-thiolate complexes to ultrabright Au(0)@Au(I)-thiolate core-shell nanoclusters. *J. Am. Chem. Soc.* **2012**, *134*, 16662–16670. [CrossRef] [PubMed]
32. Deng, H.H.; Shi, X.Q.; Wang, F.F.; Peng, H.P.; Liu, A.L.; Xia, X.H.; Chen, W. Fabrication of water-soluble, green-emitting gold nanoclusters with a 65% photoluminescence quantum yield via host-guest recognition. *Chem. Mater.* **2017**, *29*, 1362–1369. [CrossRef]
33. Xie, M.; Wang, Y.; Liu, L.; Wang, X.; Jiang, H. Luminescent gold-peptide spheric aggregates: Selective and effective cellular targeting. *J. Colloid. Interface Sci.* **2022**, *614*, 502–510. [CrossRef] [PubMed]
34. Xia, M.; Sui, Y.; Guo, Y.; Zhang, Y. Aggregation-induced emission enhancement of gold nanoclusters in metal-organic frameworks for highly sensitive fluorescent detection of bilirubin. *Analyst* **2021**, *146*, 904–910. [CrossRef] [PubMed]
35. Jia, H.; Yang, L.; Dong, X.; Zhou, L.; Wei, Q.; Ju, H. Cysteine modification of glutathione-stabilized Au nanoclusters to red-shift and enhance the electrochemiluminescence for sensitive bioanalysis. *Anal. Chem.* **2022**, *94*, 2313–2320. [CrossRef] [PubMed]
36. Hui, S.; Liu, Q.; Huang, Z.; Yang, J.; Jiang, S. Gold nanoclusters-decorated zeolitic imidazolate frameworks with reactive oxygen species generation for photoenhanced antibacterial study. *Bioconjug. Chem.* **2020**, *31*, 2439–2445. [CrossRef]
37. Feng, A.; Jiang, Q.; Song, G.; Xu, Z.; Liu, X. DNA-templated NIR-emitting gold nanoclusters with peroxidase-like activity as a multi-signal probe for Hg^{2+} detection. *Chin. J. Anal. Chem.* **2022**, *50*, 16662–16670. [CrossRef]
38. Ravichandran, M. Interactions between mercury and dissolved organic matter-a review. *Chemosphere* **2004**, *55*, 319–331. [CrossRef]
39. Yan, F.; Shi, D.; Zheng, T.; Yun, K.; Zhou, X.; Chen, L. Carbon dots as nanosensor for sensitive and selective detection of Hg^{2+} and L-cysteine by means of fluorescence "Off-On" switching. *Sens. Actuators B Chem.* **2016**, *224*, 926–935. [CrossRef]
40. Li, X.; Qiao, J.; Li, Z.; Qi, L. Fluorescence turn-off-on for highly selective detection of serum l-cysteine based on AuNCs-AuNPs ensembles. *Analyst* **2020**, *145*, 2233–2237. [CrossRef]
41. Yu, H.; Liu, Y.; Wang, J.M.; Liang, Q.; Liu, H.; Xu, J.; Shao, S.J. A gold nanocluster-based ratiometric fluorescent probe for cysteine and homocysteine detection in living cells. *N. J. Chem.* **2017**, *41*, 4416–4423. [CrossRef]
42. Wang, Y.W.; Tang, S.; Yang, H.H.; Song, H. A novel colorimetric assay for rapid detection of cysteine and Hg^{2+} based on gold clusters. *Talanta* **2016**, *146*, 71–74. [CrossRef] [PubMed]
43. Zhang, S.; Lin, B.; Yu, Y.; Cao, Y.; Guo, M.; Shui, L. A ratiometric nanoprobe based on silver nanoclusters and carbon dots for the fluorescent detection of biothiols. *Spectrochim. Acta. A Mol. Biomol. Spectrosc.* **2018**, *195*, 230–235. [CrossRef] [PubMed]
44. Lian, J.; Liu, P.; Jin, C.; Liu, Q.Y.; Zhang, X.; Zhang, X. Flower-like CeO_2/CoO p-n heterojuncted nanocomposites with enhanced peroxidase-mimicking activity for L-cysteine sensing. *ACS Sustain. Chem. Eng.* **2020**, *8*, 17540–17550. [CrossRef]

biosensors
MDPI

Article

A Combination of Near-Infrared Hyperspectral Imaging with Two-Dimensional Correlation Analysis for Monitoring the Content of Alanine in Beef

Fujia Dong [1], Yongzhao Bi [1], Jie Hao [1], Sijia Liu [1], Yu Lv [1], Jiarui Cui [1], Songlei Wang [1,*], Yafang Han [1] and Argenis Rodas-González [2]

1 School of Food and Wine, Ningxia University, Yinchuan 750021, China
2 Department of Animal Science, Faculty of Agricultural and Food Sciences, University of Manitoba, Winnipeg, MB R3T 2N2, Canada
* Correspondence: wangsonglei@nxu.edu.cn

Abstract: Alanine (Ala), as the most important free amino acid, plays a significant role in food taste characteristics and human health regulation. The feasibility of using near–infrared hyperspectral imaging (NIR–HSI) combined with two–dimensional correlation spectroscopy (2D–COS) analysis to predict beef Ala content quickly and nondestructively is first proposed in this study. With Ala content as the external disturbance condition, the sequence of chemical bond changes caused by synchronous and asynchronous correlation spectrum changes in 2D–COS was analyzed, and local sensitive variables closely related to Ala content were obtained. On this basis, the simplified linear, nonlinear, and artificial neural network models developed by the weighted coefficient based on the feature wavelength extraction method were compared. The results show that with the change in Ala content in beef, the double-frequency absorption of the C-H bond of CH_2 in the chemical bond sequence occurred prior to the third vibration of the C=O bond and the first stretching of O-H in COOH. Furthermore, the wavelength within the 1136–1478 nm spectrum range was obtained as the local study area of Ala content. The linear partial least squares regression (PLSR) model based on effective wavelengths was selected by competitive adaptive reweighted sampling (CARS) from 2D–COS analysis, and provided excellent results (R^2_C of 0.8141, R^2_P of 0.8458, and RPDp of 2.54). Finally, the visual distribution of Ala content in beef was produced by the optimal simplified combination model. The results show that 2D–COS combined with NIR–HSI could be used as an effective method to monitor Ala content in beef.

Keywords: hyperspectral imaging; two–dimensional correlation spectroscopy; alanine; visualization

check for updates

Citation: Dong, F.; Bi, Y.; Hao, J.; Liu, S.; Lv, Y.; Cui, J.; Wang, S.; Han, Y.; Rodas-González, A. A Combination of Near-Infrared Hyperspectral Imaging with Two-Dimensional Correlation Analysis for Monitoring the Content of Alanine in Beef. *Biosensors* **2022**, *12*, 1043. https://doi.org/10.3390/bios12111043

Received: 22 October 2022
Accepted: 15 November 2022
Published: 18 November 2022

1. Introduction

Beef is a popular food in the human diet, and its rich amino acid and protein composition is close to human needs [1]. With the rapid development of social economy and the continuous improvement in life quality, meat quality plays an increasingly important role in determining the value of meat products, and more and more consumers are attracted to high-quality meat [2]. As the most critical factor affecting the quality of meat products, flavor not only affects the taste of food and the absorption of nutrients, but also determines the consumers' purchase desire and intake intention to a certain extent [3]. Free amino acids (FAA) are the main taste substances in meat products. Their type and content have an important impact on meat quality, antioxidant activity, and nutritional value. They can also be used as precursors for the Maillard reaction and Strecker degradation reaction with reducing sugar, affecting the overall flavor of the food system [4]. Different FAAs have different taste characteristics. They have a low taste threshold, strong taste ability, and five basic taste senses: sour, sweet, bitter, salty, and umami [5]. Among them, alanine (Ala), as the most simple-flavored amino acid among FAAs, has become the main sweet amino acid in meat products due to its low hydrophobicity, and the umami of food will be directly affected by its content [6]. When Ala coexisted with taste substances such as glutamic acid and ornithine in food, it can produce a synergistic effect and provide strong umami for meat products [7]. In addition, Ala also plays a variety of important physiological roles, including improving the immune system, preventing and treating vascular diseases, and participating in growth and metabolism [8]. When too much or too little Ala is ingested, the absorption balance of human total amino acids might be affected, leading to nutritional imbalance and poor health [9]. Therefore, it is of great significance to develop a rapid, nondestructive, and noncontact quantitative method for the determination of Ala content in beef.

At present, high-performance liquid chromatography and automatic amino acid analysis are often used for the physical and chemical detection of FAA [2–4]. These methods have the advantage of high precision in detecting the composition and content of amino acids. However, their disadvantage is that sample pretreatment is complex, harmful, and polluting, and the integrity of the sample is damaged [10]. Therefore, the rapid detection requirements in the beef mass production process cannot be met. In previous investigations, the combination of several nondestructive rapid measurement methods and chemometric methods have been applied in the assessment of amino acid content, including visible near–infrared spectroscopy, near–infrared (NIR) spectroscopy, Fourier infrared spectroscopy, and nondestructive magnetic resonance imaging [11–14]. However, these studies mainly focused on the evaluation of research objectives concerning soybean, daqu, tea, potato, and ham [11–15], and detection indicators such as amino acid nitrogen [12], total amino acid [15], and total volatile basic nitrogen (TVB-N) have been emphatically discussed [16]. In addition, hyperspectral imaging (HSI) technology is more widely focused in predicting other meat-related quality attributes, especially nutritional attributes (fatty acid, protein, and intramuscular fat), technical attributes (pH and water holding capacity), sensory attributes (tenderness, color, hardness, gumminess, and chewiness), freshness attributes (thiobarbituric acid reactive substances (TBARS), total biogenic amines (TBA), and myoglobin), and microbial attributes (total viable count) of meat in different parts, types, and places of origin [17–25]. Notably, Cheng et al. [23,24] reported that NIR–HSI had great application potential in evaluating the content of meat quality (TBA, TBARS, and fat oxidation). Through comparison, it was found that molecules with greater contributions could be detected more easily than those with smaller contributions. The above research provides the possibility to reveal NIR–HSI prediction of Ala content in meat and meat products.

However, beef, as a complex food item, has interactions among various components (proteins, amino acids, lipids, and carbohydrates) [26]. This makes it difficult to extract the NIR spectral information of the meat, and the spectral signal presents overlapping and complexity [27]. As a result, quantitative analysis by NIR has relied heavily on the application of chemometric analysis to relate the subtle spectral changes to the variations in concentrations of certain components in the analyte [28]. In previous studies, the derivative method, weight value method, principal component analysis, and feature variable extraction were the main spectral response analysis methods for studying the spectral characteristic variables of objects [29,30]. It usually requires a lot of experiments to identify sensitive variables and build a stable model effect. This is a time-consuming and iterative process, the results of which vary with experience and the chemometric methods used. Thus, a better understanding of NIR spectra and a more accurate spectral band division is conducive to the establishment of a more robust NIR quantitative model. In recent years, two–dimensional correlation spectroscopy (2D–COS) has been applied to HSI research by some researchers to improve spectral resolution by extending one–dimensional spectral signals to the second dimension [31]. By this means, the changes in subtle spectral features are analyzed and the relationship and change order of various groups are revealed [32]. This technique has been successfully applied to spectral interpretation and spectral band allocation of the lipid oxidation of meat, the damage of myofibrils, and the spectral interpretation and distribution of protein secondary structure changes in meat [24,26]. In addition, Dong et al. [33] reported that the use of 2D–COS to select the NIR–HSI continuous sensitive interval and the establishment of a deep learning algorithm have great potential to improve the accuracy of the model. This provides a new direction for our research. As far as we know, a feasibility study using NIR–HSI combined with 2D–COS analysis to detect Ala content in beef has not been previously reported.

Therefore, this study is the first to explore the feasibility of NIR–HSI combined with 2D–COS analysis in detecting Ala content in beef. The specific objectives were as follows: (i) NIR–HSI (900–1700 nm) was used to collect the spectral information of beef samples, the segment threshold method was used to select the sample region of interest (ROI), and the Monte Carlo (MC) method was used to eliminate abnormal value information; (ii) the determination of the change order of characteristic peaks related to Ala content and local sensitive intervals was achieved by analyzing synchronous and asynchronous 2D–COS; (iii) the determination of the best characteristic variables of the global and local spectral intervals based on the weight algorithm (competitive adaptive reweighted sampling (CARS) and regional coefficient (RC)) were studied; (iv) simplified linear partial least squares regression (PLSR), nonlinear least squares support vector machine (LSSVM), and artificial neural network (ANN) Ala prediction models were developed; (v) the optimal characteristic variables and models obtained were used to characterize the visual distribution of Ala content. The research results were expected to further improve the accuracy of NIR–HSI technology in detecting meat quality indicators. A graphical representation of the proposed method is illustrated in Figure 1.

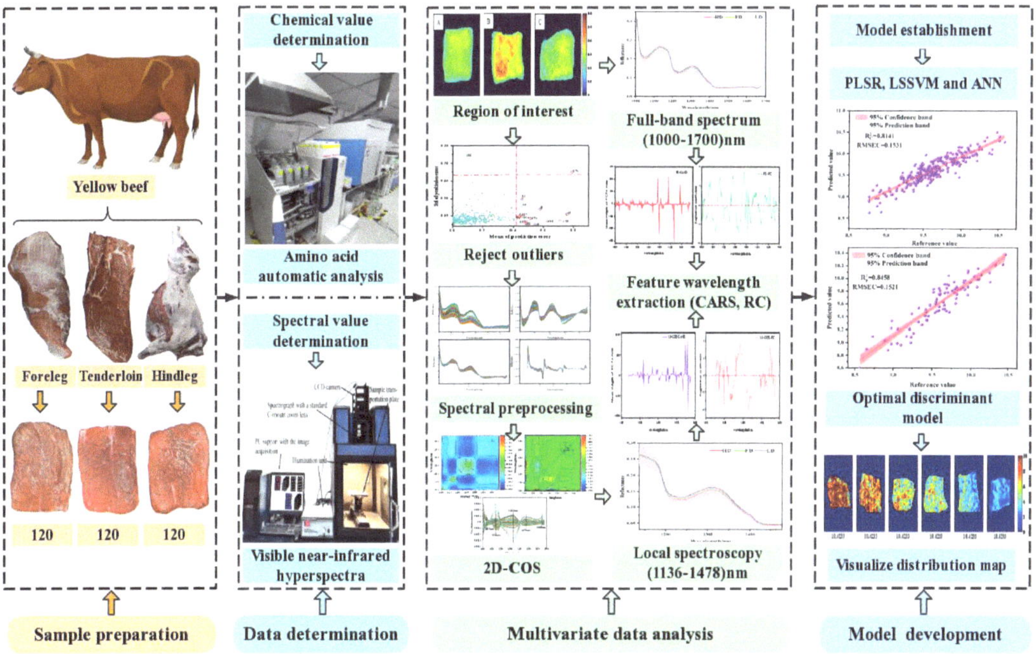

Figure 1. Technical roadmap of NIR–HSI detection of Ala content.

2. Materials and Methods

2.1. Sample Preparation

The samples were collected from three parts (longissimus dorsi (LD), foreleg (FD), and hind leg (HD)) of 20 cattle in Ningxia, China. The samples were vacuum packed and stored in a portable refrigerated incubator, and were transported to the Meat Processing and Quality Safety Control Laboratory of Ningxia University. The oil and fascia in the fresh samples were removed and cut into 40 mm × 40 mm × 20 mm (L × W × T). All samples were vacuum packed and stored in a 4 °C refrigerated room. In order to ensure the reliability and universality of the model, the samples were collected following four slaughter batches (30 samples were obtained from each part of each batch). Finally, 360 beef samples were obtained, and their spectral data and chemometric data were measured at the same time.

2.2. *Hyperspectral Image Correction and Parameter Determination*

The NIR–HSI (900–1700 nm) system was used to collect spectral images of beef samples. The HSI system can continuously acquire 256 spectral bands with a spectral resolution of 5 nm. The HSI system was mainly composed of five parts, including an HSI spectrometer, four 35 W tungsten halogen lamps, a CCD camera, an electronic displacement platform, and a computer. The best debugging parameters were optimized during preliminary experiments because diffuse reflections of the light source may be caused by the color, texture, and shape of the beef sample. The best debugging parameters were as follows: the object distance was 360 mm, the steady current of the light source was set to 6.0 A, the electric control displacement speed was 20 mm/s, and the exposure time of the camera was 30 ms. In order to reduce the uneven distribution of image light source intensity and the existence of a dark current in the sensor, black and white correction was required for the obtained hyperspectral image. The formula used was as follows:

$$R = \frac{A - S}{B - S} \times 100\% \quad (1)$$

where A is the original spectral image of the sample; B is the all-white calibration image; S is the all-black calibration image; R is the calibrated spectral image. The all-black calibration image S was obtained by covering the camera lens (almost 0% reflectance), and the all-white calibration image B was obtained using a white board made of polytetrafluoroethylene (>99% reflectance). The ROI of the HSI was extracted from the spectral information of the sample using the segmented threshold method (set at 0.16) with ENVI software.

2.3. *Measurement of the Content of FAAs*

Sample pretreatment: minced meat sample (2.00 g) was weighed, 0.02 mol/L hydrochloric acid was added, and the sample was then placed in a 10 mL centrifuge tube for homogenization. After ultrasound (30 min), centrifugation (4000 r/5 min), and activation (C18), 5.00 mL methanol and 5.00 mL water were added, respectively. After filtration, 2.5 mL of solution was absorbed and 1.50 mL hydrochloric acid of 0.02 mol/L was added. After passing through the column, 0.02 mol/L hydrochloric acid was used to dilute the solution to 5.00 mL. After uniform mixing, the solution was centrifuged (10,000 r/10 min) after standing (15 min), and then filtered through a membrane of 0.45 μm pore size for analysis. By comparing the retention time and peak area of each amino acid standard, qualitative and quantitative analysis of each amino acid was carried out.

Analytical parameters: the chromatographic column used was a sulfonic acid cation resin separation column (4.6 mm × 60 mm). The detection wavelength was 440 nm and 570 nm, respectively; the injection volume was 20 μL. The reaction temperature was 135 ± 5 °C. The separation column temperature was 57 °C.

2.4. *Analysis of Two–Dimensional Correlation Spectra*

As an advanced spectral analysis method, 2D–COS is particularly suitable for exploring the structural changes and interactions of complex systems under external disturbances from the molecular perspective [31]. Compared with traditional one–dimensional spectral analysis, 2D–COS has a strong simplification effect for complex spectra containing multiple overlapping peaks. At the same time, it is extended on the basis of the original one–dimensional spectrum, significantly improving the resolution of the original spectrum [32].

With the help of a peak correlation diagram, the assignment and interaction of peaks in the system were judged, and the change order of peak position under external disturbance was obtained [26]. In this study, Ala was used as the external disturbance condition, and the dynamic spectrum $\tilde{y}\ (v,\ d)$ caused by the system in the external disturbance range (1~T) was defined as:

$$\tilde{y}(v,d) = \begin{cases} y(v,d) - \overline{y}(v)\ 1 \leq d \leq T \\ 0\ other\ wise \end{cases} \tag{2}$$

where $\overline{y}\ (v)$ is the reference spectrum of the system, which was usually set as the average spectrum, and was defined as:

$$\overline{y}(v) = \frac{1}{T}\sum_{j=1}^{T} y(v,d_j) \tag{3}$$

where $\overline{y}\ (v,\ d)$ data are expressed in discrete form in actual measurement. The following vector forms were commonly used:

$$\overline{y}(v,d) = \begin{bmatrix} y = (v,d_1) \\ y = (v,d_2) \\ y = (v,d_3) \\ \cdot \\ \cdot \\ \cdot \\ y = (v,d_m) \end{bmatrix} \tag{4}$$

The two–dimensional correlation intensity $X\ (v_1,\ v_2)$ indicated the function of the spectral intensity changes $\tilde{y}\ (v,\ d)$ of the spectral variables v_1 and v_2, and was compared in the external disturbance variable interval. The correlation function was used to calculate the intensity change at two independent spectral variables, v_1 and v_2, so that $X\ (v_1,\ v_2)$ could be converted into the plural form:

$$X(v_1,v_2) = \Phi(v_1,v_2) \cdot i\,\Psi(v_1,v_2) \tag{5}$$

According to the 2D–COS theory of Noda et al. [31,32], the mutually perpendicular real parts and imaginary parts of the complex number were called the synchronous correlation strength and asynchronous correlation strength, respectively, and the strength changes of the two were directly related to the change in d value. We then converted the dynamic spectrum from the external interference domain to the frequency domain via Hilbert–Noda change, and finally, 2D–COS was obtained. Its two–dimensional correlation synchronous spectrum was expressed as (6):

$$\Phi(v_1,v_2) = \frac{1}{T-1}\tilde{y}(v_1)^T \cdot \tilde{y}(v_2) \tag{6}$$

The expression of the two–dimensional correlation asynchronous spectrum was (7):

$$\Psi(v_1,v_2) = \frac{1}{T-1}\tilde{y}(v_1)^T \cdot \mathrm{N} \cdot \tilde{y}(v_2) \tag{7}$$

where N is the T-order square matrix (T is the spectral number), which was called the Hilbert–Noda matrix; the matrix formula was (8):

$$N_{jk} = \begin{cases} o_j = k \\ \frac{1}{\pi(k-j)}\ j \neq k \end{cases} \tag{8}$$

2.5. Analysis Rules of Spectral Peak

The synchronous correlation spectrum characterizes the simultaneous or coincidental changes in spectral intensities measured at spectral variables of v_1 and v_2. In the atlas, it was symmetrical along the diagonal direction, its autocorrelation peak appeared on the diagonal, and the cross peak appeared outside the diagonal. In the synchronous spectrum, the intensity of the automatic peak was always positive, representing the overall degree of the dynamic change in spectral intensity under the corresponding number of bands. It is worth noting that there were positive and negative cross peaks in the synchronous spectrum. If the cross peaks of the two bands were positive, it meant that the spectral intensity of the corresponding wave number increased or decreased simultaneously under external interference. When the opposite value was negative, it meant that the spectral intensity corresponding to the two wave numbers increased one and decreased the other. In the asynchronous two–dimensional correlation spectrogram, the asynchronous or sequential (i.e., delayed or accelerated) changes in spectral intensity at the given wave numbers v_1 and v_2 were presented. Its asynchronous graph was asymmetric with respect to the diagonal, and only had cross peaks. The asynchronous cross peak only appeared when the spectral intensity of a given wave number changed out of phase. Using this feature to analyze the overlapping peaks with different sources in the spectrum had a significant role in judging the change order of the characteristic peaks in the process of external interference. The direction and order of strength change determined according to Noda rules are shown in Table 1.

Table 1. The direction and sequence of intensity changes according to Noda's rule.

Φ (v_1, v_2)	Ψ (v_1, v_2)	Significance
+	/	The signal strength at v_1 and v_2 changes in the same direction, i.e., increases or decreases at the same time.
−	/	The signal strength at v_1 and v_2 changes in opposite directions.
+	+	The change at v_1 is mainly prior to the change in band at v_2.
+	−	The change at v_1 mainly follows the change in wave band at v_2.
−	+	The change at v_1 mainly follows the change in wave band at v_2.
−	−	The change at v_1 is mainly prior to the change in band at v_2.

Note: Φ (v_1, v_2), synchronous correlation spectrum; Ψ (v_1, v_2), asynchronous correlation spectrum.

2.6. Extraction of Spectral Characteristic Wavelength

The complex and time-consuming properties of model training were caused by the high dimension of the data and the strong correlation between adjacent variables. Therefore, the selection of characteristic wavelength variables became a key step in spectral analysis, which was mainly used to simplify the model and eliminate data redundancy. The use of genetic algorithms, principal component algorithms, iterative algorithms, and weight algorithms to extract feature variables has been reported by a large number of researchers [27–29]. In this study, two weight algorithms (CARS and RC) were used to select characteristic wavelengths to develop simplified models for the full spectral (FS) area and sensitive local areas selected from 2D–COS analysis. The purpose of this was to consider the proportion of data from the perspective of weight to analyze the appearance of characteristic variables more reasonably.

2.7. Visualization of the Ala Contents

Inversion of Ala content distribution was feasible because HSI had both image information and spectral information (combining two–dimensional imaging technology with one–dimensional spectral curve to form three–dimensional data). Based on the method of multivariate optimization model, the optimal simplified combination was selected, the weight coefficients of each pixel and the optimal correction model were calculated, and a matrix consisting of multiple predicted values was obtained. Then, the obtained matrix was refolded to generate the content distribution map [30]. The Ala content of each pixel was expressed in different color scales. Therefore, the distribution of Ala content could be clearly inverted on the color map.

2.8. Model Establishment and Evaluation

In this study, three multivariate algorithms, including linear PLSR, nonlinear LSSVM, and ANN, were used to develop the detection model for quantitative analysis. PLSR is a multiple factor regression method. Firstly, the scores of the main factors were extracted from the spectral matrix X and the physical and chemical matrix Y, and the PLS was used to conduct the best precision regression for the main factors of X and Y, respectively. The principal component of spectral matrix X was directly related to physical and chemical parameters, and the linear relationship between spectral variables and physical and chemical parameters was used to the greatest extent. LS-SVM used the least squares linear equation as the loss function formula. The convex quadratic programming was solved by solving linear equations instead of traditional SVM, which reduced the training time and computational complexity. ANN is a multilayer feedforward neural network characterized by forward signal propagation and backward error propagation. According to the error signal of forward propagation, the method of gradient descent was adopted for backward propagation, and the signal error was minimized through repeated forward and backward learning.

In order to establish the reliable accuracy of the validation model, the whole sample set data were divided according to a 3:1 ratio, based on the RS method. The proficiency and accuracy of the model were evaluated by analyzing statistical parameters. These included determination coefficients (R^2), root mean square error (RMSE), and ratio of performance deviation (RPD). Generally, a good model should have higher R^2 and RPD values and lower RMSE values.

3. Results and Discussion

3.1. Spectral Reflectance Index Visualization and Spectral Curve Analysis

ENVI software was used to extract sample ROI from HSI by the segmentation threshold method, and to visually express spectral reflectance indexes of different parts. First, the band with the best image definition was selected, and the optimal segmentation threshold of samples and background was set to 0.16 to retain the effective sample area as much as possible and eliminate the influence of background area interference. The independent regions of each sample were taken as ROI, and their average values were taken as the effective spectral data of the sample. The visual distribution of spectral reflectance index of beef samples from different parts is shown in Figure 2. When the color was closer to red, it meant that the reflectivity index was larger, and vice versa. It can be observed that the reflectivity index of the FD and the HD was low, and their colors were similar to each other, showing a yellow-green coloration, while the LD appeared an obvious yellow color with a small amount of red, and the reflectivity index was large. There was a certain correlation between the spectral reflectance index and the chemical composition of the sample, indicating that the LD had a more complex chemical composition than the FD and HD.

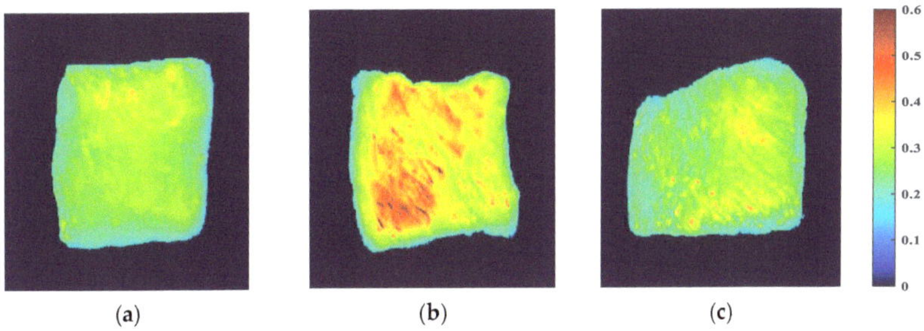

Figure 2. Visualization of spectral reflectance index of beef samples from different parts: (**a**) FD, (**b**) LD, and (**c**) HD.

The spectral curves of 360 beef samples obtained by NIR–HSI are shown in Figure 3. The NIR bands of the beef samples contained rich combination bands of molecular overtones and molecular vibration, which were characterized by the double-frequency absorption of chemical bonds and relatively strong spectral characteristic absorption. As shown in Figure 3a, it was observed that there was more noise in the full-band original spectral curve of beef samples within the band range of 918–1000 nm, which might be related to the instability of the instrument at the beginning of scanning. Therefore, the spectral data of 1000–1700 nm (225 wave bands) were further analyzed. Figure 3b shows the average spectral curves of different beef samples. It can be seen that the curves show a similar trend as a whole. The spectral curve of the LD muscle was higher than that of the FD and HD, which is basically consistent with the conclusion drawn from the visualization of the ROI spectral reflectance index of the sample. It is worth noting that the longitudinal shifts of the spectra were rather different. Specifically, compared with the spectral reflectance of the 1010–1400 nm band, the reflectance of the 1400–1700 nm band was low and the peak amplitude frequency was wide. This difference might be related to the typical reflection characteristics of different components in the sample at a specific wave band when the electromagnetic radiation wave interacted with the material.

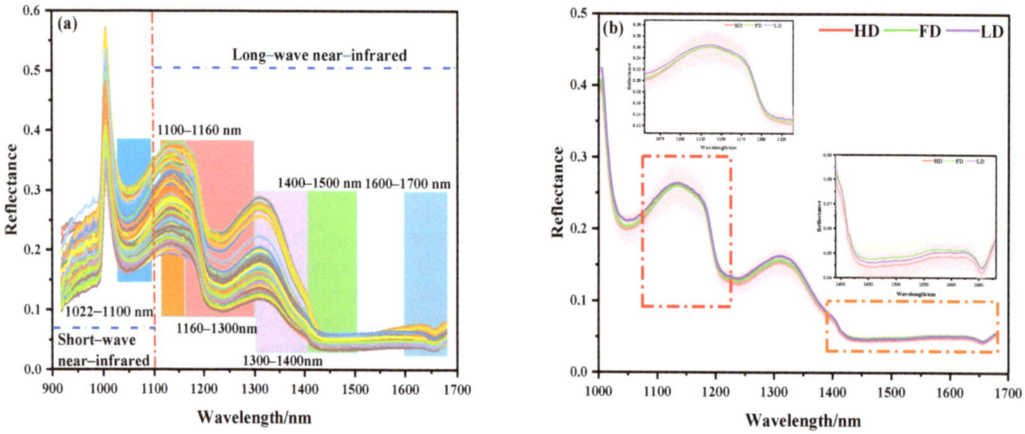

Figure 3. NIR spectral curves of beef samples: (**a**) full spectral curve, (**b**) spectral curves of different parts.

According to relevant studies, the spectral absorption band in the NIR band was mainly related to the tensile vibration of amide bonds, C-H bonds, N-H bonds, and O-H bonds of organic compounds (proteins and amino acids), lipids, carbohydrates, and water [33]. The absorption band at 1022–1100 nm might be related to the second stretching of the N-H bond in amides. The peak at 1100–1160 nm might be related to the double-frequency absorption of the C-H bond of the CH_2 group. The absorption band at 1160–1300 nm was attributed to the second C-H stretching of carbonyl compounds. The absorption band at 1300–1400 nm represents the third vibration the of C=O bond. The absorption band at 1400–1500 nm represents the first O-H stretching of COOH. The infrared spectrum in the 1600–1700 nm region was related to the amide I band of the protein, which was mainly attributed to the absorption of the C=O bond [17,22,30].

3.2. Abnormal Sample Detection and Sample Set Division

As a common method to eliminate outliers, MC can help identify the variation among samples and improve the accuracy of the model. It was mainly used to calculate the corresponding mean value and standard deviation, via a random sampling method, and to further draw the distribution map. Generally, the points far away from the main sample were assumed to be outliers, and were thus eliminated. Abnormal samples were identified based on the improvement or decrease in the performance of the established PLSR model. As shown in Figure 4, 16 abnormal value samples of 360 samples (3, 4, 8, 64, 43, 56, 61, 75, 122, 168, 174, 179, 227, 273, 298, 299) were identified by MC. The modeling effect of each abnormal value was obtained by eliminating samples one by one, as shown in Table 2. When samples 64, 168, and 299 were removed, the prediction performance of the model was reduced; therefore, false samples were retained. When the remaining 13 samples were removed, the final PLSR model was established. The R^2_{cv} value increased from 0.6905 to 0.7419, and the RMSECV value decreased from 0.2071 mg/100 g to 0.1831 mg/100 g.

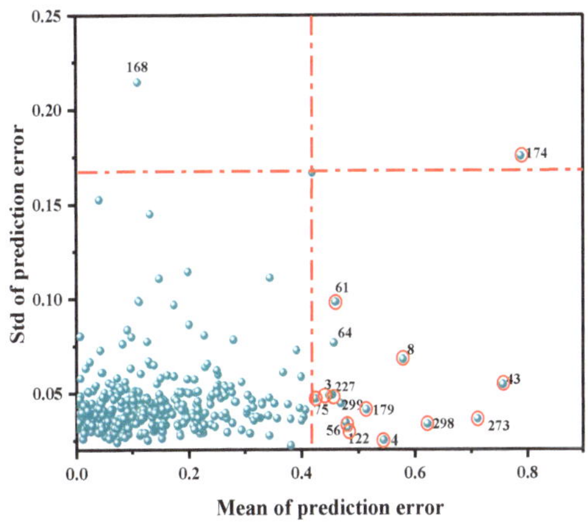

Figure 4. Detection results of abnormal Ala content in beef at the NIR band.

Table 2. Detection results of abnormal beef samples based on the Monte Carlo method.

Sample Set	Outliers	Remaining Amount	PB		PA	
			R^2_{CV}	RMSECV	R^2_{CV}	RMSECV
Ala	64, 168, 299	357	0.6905	0.2071	0.6209	0.2275
	3, 4, 8, 43, 56, 61, 75, 122, 174, 179, 227, 273, 298	347	0.6905	0.2071	0.7419	0.1831

Note: PB, performance before removing the outliers; PA, performance after removing the outliers.

The partition of sample set was an important step in spectral multivariate analysis. After removing the outliers, 260 samples were selected as the calibration set using the RS algorithm, and 87 samples were used to predict the model performance. The statistical results are shown in Table 3. It can be seen that the range of the Ala prediction set was included in the correction set, and the difference between the average value and the standard deviation of the two datasets was not large, which means that the distribution of the divided sample sets was similar. This was highly beneficial to the establishment of a prediction model with high accuracy and robustness. At the same time, it also shows that it was feasible to use the RS method to divide samples.

Table 3. Statistical results of beef sample set divided by the RS method.

Sample Set	Calibration Set					Prediction Set				
	N	**Range**	**Mean**	**SD**	**TAV**	**N**	**Range**	**Mean**	**SD**	**TAV**
Ala	260	5.04–10.6	9.616	0.356	0.160	87	5.56–10.5	9.611	0.387	0.160

Note: SD, standard deviation; TAV, taste activity value.

3.3. Analysis of Spectral Full Band Modeling

In order to maximize the resolution of overlapping data and reduce the system noise caused by spectral scattering and instrument drift, nine mathematical preprocessing methods were used to correct the full spectrum data. The full wavelength data that was relevant to Ala content were evaluated and compared according to the PLSR model (Table 4). The original spectral results show a good prediction model effect ($R^2_C = 0.8202$, RMSEC = 0.1507 mg/100 g; $R^2_P = 0.8145$, RMSEP = 0.1663 mg/100 g). The modeling performance of the nine pretreatments had a small overall difference. Compared with the original spectrum, the PLSR model established by MF pretreatment showed the highest correlation coefficient when evaluating the Ala content, i.e., R^2_C increased to 0.833, R^2_P increased to 0.8388, and RMSEP decreased to 0.1548. As a nonlinear signal processing technology, MF had a good filtering effect on spectral images. At the same time, it was able to protect the edge data of the signal. Fan et al. [27] also used MF to optimize the original NIR spectrum to determine the TBA content in Tan mutton, and obtained Rc and Rp values of 0.94 and 0.88, respectively.

Table 4. PLSR model performance of different pretreatment methods.

Sample Set	Pretreatment Method	LVs	Calibration Set		Cross-Validation		Prediction Set	
			R^2_C	**RMSEC**	**R^2_{CV}**	**RMSECV**	**R^2_P**	**RMSEP**
Ala	None	15	0.8202	0.1507	0.7527	0.1773	0.8145	0.1663
	MA	15	0.8156	0.1526	0.7570	0.1756	0.8230	0.1850
	GF	15	0.8180	0.1516	0.7561	0.1760	0.8287	0.1674
	MF	**16**	**0.8330**	**0.1452**	**0.7619**	**0.1738**	**0.8388**	**0.1548**
	SG	15	0.8129	0.1537	0.7530	0.1771	0.8191	0.1811
	Normalize	14	0.8051	0.1569	0.7510	0.1777	0.7898	0.1764
	Baseline	18	0.8136	0.1534	0.6935	0.1991	0.8165	0.1651
	SNV	13	0.7729	0.1693	0.7059	0.1934	0.7705	0.1862
	DT	17	0.8052	0.1568	0.6904	0.1997	0.8084	0.1687
	MSC	12	0.7597	0.1742	0.6941	0.1941	0.7476	0.1939

Note: Bold indicates optimal model effect. MA, moving average; GF = Gaussian filter; MF, median filter; SG, Savitzky–Golay; SNV, standard normal variate; DT, detrending; MSC, multiplicative scatter correction.

3.4. 2D–COS Analysis of Ala Content in NIR–HSI

The pretreated NIR with Ala content as the external disturbance condition was analyzed by 2D–COS, as shown in Figure 5a. The self-peak in the synchronous spectrum was positive on its diagonal. The strength of the automatic peak was reflected by the number of circles. The more circles, the stronger the automatic peak. On the contrary, the fewer circles present, the weaker the automatic peak. The corresponding cut spectrum in Figure 5c was generated in order to more clearly represent the position of its self-peak. It can be seen that five main self-peaks were 1055, 1136, 1323, 1478, and 1648 nm at the diagonal position. The appearance of these autocorrelation peaks indicates that with the change in Ala content, the spectral absorption intensity of the band underwent a strong change, and the spectral signals of these characteristic variables were more sensitive to external interference. Among them, 1055 nm belonged to the second stretching of the N-H bond in amides, 1136 nm was the double-frequency absorption band of the C-H bond in CH_2, 1323 nm was the third vibration of the C=O bond, 1478 nm was the first O-H stretching of COOH, and 1648 nm was the absorption stretching of the C=O bond in the amide I band of protein [22–30].

Analyzing the cross peaks outside the main diagonal of the synchronous spectrum, only the obvious cross peaks were observed at (1136–1323), (1323–1478), and (1478–1136) nm. However, there was no cross peak between (1055–1136) and (1478–1648) nm, indicating that the absorption peak between them was irrelevant; thus, the self-peak at 1055 nm and 1648 nm was excluded. At the same time, it was found that the cross peaks generated by the 1136 nm and 1478 nm (named band group 1) bands were always positive, indicating that the spectral intensity of these wavelengths increased or decreased with the fluctuation of disturbance. The cross peak generated in the 1323 nm (named band group 2) band was always negative, which indicates that the bands from group 1 and group 2 demonstrated opposite behaviors.

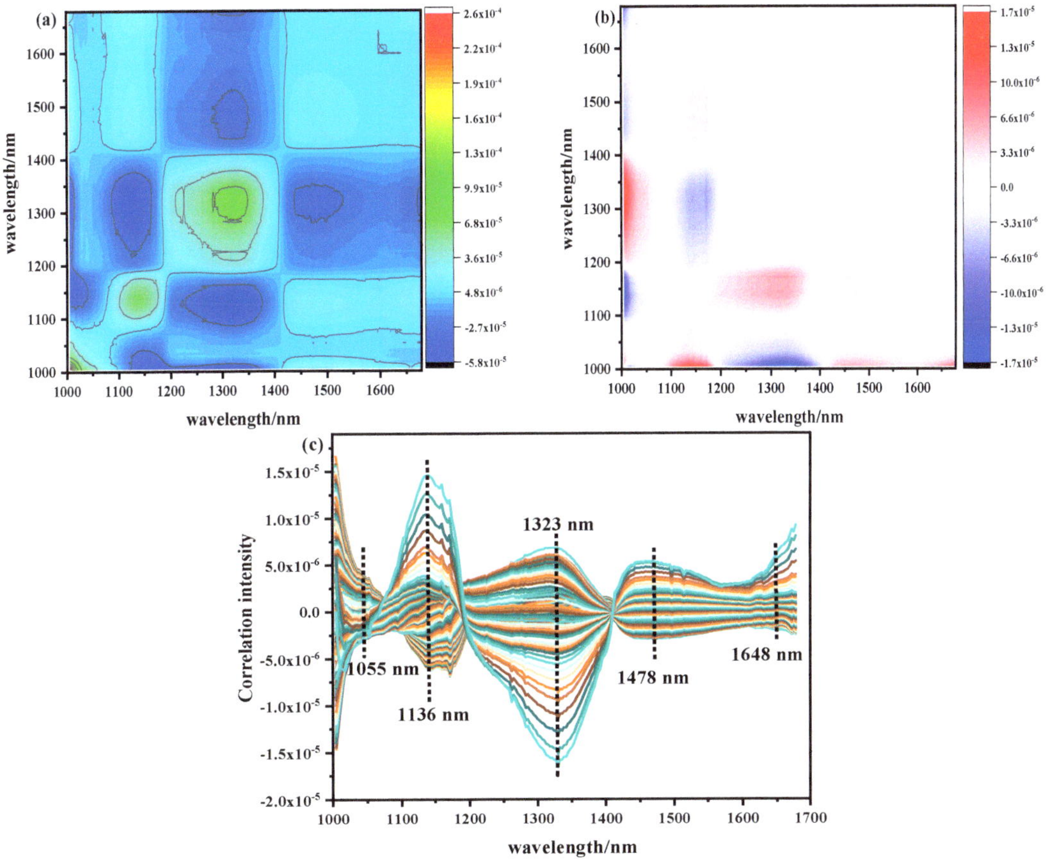

Figure 5. NIR two–dimensional correlation spectrum and cutting spectrum of Ala content: (**a**) synchronous two–dimensional correlation spectrum, (**b**) asynchronous two–dimensional correlation spectrum, (**c**) two–dimensional correlation cut spectrum.

The asynchronous 2D–COS is shown in Figure 5b. When the phase of the spectral intensity change in two variables was different, cross peaks were generated and used to infer the order of spectral intensity change. The same main cross peak was observed at the self-peak position of the synchronous spectrum. The signs of the cross peaks are shown in Table 5. According to the positive and negative signs of synchronous asynchronous cross peaks, it can be concluded that under the condition of Ala content as an external disturbance, the order of changes in the relevant spectra was 1136–1323–1478 nm. This shows that under the change in Ala content in beef, the double-frequency absorption of the C-H bond (1136 nm) in CH_2 occurred before the third vibration of the C=O bond (1323 nm) and the first O-H stretching of COOH (1478 nm).

Table 5. Cross peak sign of two–dimensional correlation spectrum of Ala content in the NIR band.

Wavelength/nm	1136	1323	1478
Assignment	C-H	C=O	O-H
Synchronous			
1136	+	−	+
1323		+	−
1478			+
Asynchronous			
1136	\	−	+
1323		\	−
1478			\

From the above analysis, we could see that more sensitive variables could be obtained by 2D–COS compared with the one–dimensional spectral curve. Therefore, in the further study, the sensitive range 1136–1478 nm (115 wave bands) of 2D–COS analysis was selected as the detection area of beef Ala content. Its local region contained the main absorption band of the one–dimensional spectrum. The spectral resolution was significantly improved by this method, and the spectral band contributions from different sources and the order of chemical bond changes were distinguished.

3.5. Characteristic Wavelength Extraction

In order to improve the prediction performance of the model to a certain extent, the feature variable screening method based on the weight algorithm was selected to evaluate the global data and local regional data of Ala. The purpose of this was to verify the impact of different weighting methods on the model effect [34].

As a fast variable selection method, CARS algorithm was proposed according to the principle of "survival of the fittest" in Darwin's theory of evolution. Due to the randomness of the MC sampling method in the CARS algorithm, in order to ensure the reliability of the model, each set MC sampling number was ran 500 times, respectively, and a 10-fold cross-validation method was used to take the wavelength corresponding to the minimum RMSECV value in all PLSR models as the optimal variable [35]. Figure 6a shows the average weight distribution of CARS. It can be seen that the characteristic variables of the whole band selected by CARS were evenly distributed, and the weight value changed slightly. On the contrary, the position of characteristic variables appearing in the local band selected by 2D–COS was denser and closer to the position selected in the full band, but the weight value was significantly higher. This shows that the local features selected by 2D–COS contained a large number of features, and the correlation between features was large. In the case of local data, the disappearance of some features would not affect the detection and matching of other features [35].

There were 36 characteristic variables selected by FS–CARS, including 1029, 1038, 1068, 1115, 1130, 1136, 1151, 1163, 1199, 1217, 1235, 1238, 1246, 1267, 1288, 1336, 1339, 1378, 1381, 1384, 1395, 1410–1416, 1431, 1437, 1446, 1458, 1476, 1482, 1518, 1533, 1586, 1592, 1601, and 1661 nm. In 2D–COS–CARS, 36 characteristic wavelengths were selected, including 1136, 1139, 1187, 1193, 1199, 1205, 1211, 1220, 1240, 1246, 1267, 1288, 1297, 1303, 1309, 1321, 1342–1348, 1354, 1357, 1363, 1372, 1384, 1395, 1398, 1413–1419, 1425, 1431, 1434, 1446, 1458, 1461, and 1473 nm. We found that when feature variables were extracted by the CARS algorithm, the number of extracted variables remained stable without significant changes in the total number of bands of the model. Wan et al. [25] also reported that in the selection of characteristic wavelengths for HSI detection of OxyMb and MetMb content in Tan mutton, the CARS algorithm was used to extract 36 and 33 characteristic variables, respectively, to establish linear and nonlinear models that provided the best results. In addition, Cheng et al. and Zhuang et al. [36,37] reported that 22 and 26 characteristic variables were extracted by the CARS algorithm to obtain effective prediction results in evaluating meat DeoMb and TVB-N content. The above reports strongly demonstrate the robustness and effectiveness of the CARS algorithm, and the number of selected feature wavelengths in these studies was between 20–40, which is similar to the number of feature wavelength extractions in this study.

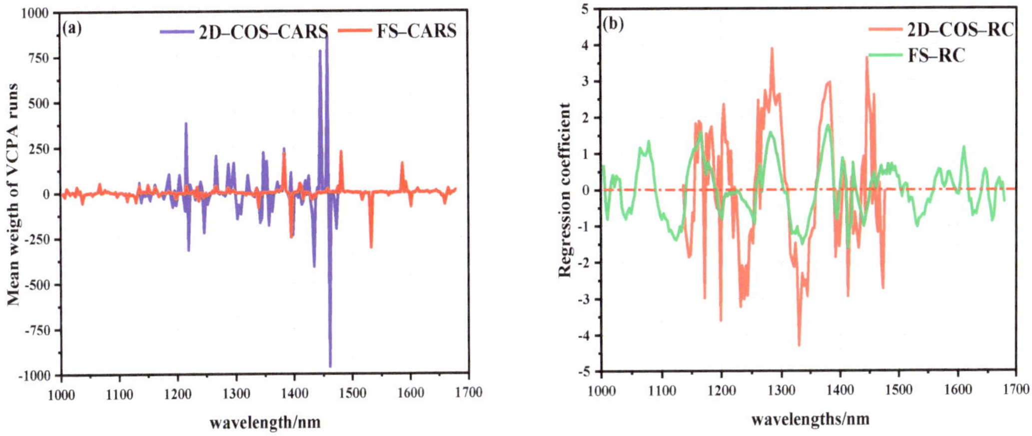

Figure 6. Operation diagram of characteristic wavelength extracted by weight algorithm: (**a**) average weight distribution diagram of CARS algorithm, (**b**) RC algorithm weight regression coefficient distribution diagram.

The RC algorithm, as a common spectral peak analysis method, is widely used in HSI nondestructive testing. The RC algorithm conducted PLSR modeling and analysis on the reflectance vector X corresponding to each wavelength in the spectral matrix of the calibration set and the component value vector Y to be measured in the physicochemical matrix, and the wavelength regression coefficient distribution map was obtained. The larger the absolute value of the regression coefficient corresponding to the wavelength point, the more information obtained, and the stronger the correlation. Therefore, the wavelength with large absolute value of regression coefficient was selected to participate in the model establishment process [34]. Figure 6b shows the distribution of weight regression coefficients for the prediction of beef Ala content in two spectral intervals using the RC algorithm. It was also found that the local weight of 2D–COS selection was significantly higher than the global weight, and there were relatively many characteristic variables in the local area with strong significant positive correlation peaks.

In FS–RC, 26 characteristic variables were selected, of which 1005, 1020, 1080, 1165, 1285, 1381, 1404, 1422, 1488, 1568, 1601, 1610, 1646, and 1670 nm were positively correlated with Ala, and 1011, 1041, 1124, 1255, 1336, 1395, 1413, 1440, 1533, 1625, 1658, and 1679 nm showed negative correlations with Ala. At the same time, 22 characteristic variables were selected in 2D–COS–RC, among which 1165, 1184, 1193, 1195, 1205, 1264, 1285, 1381, 1407, 1425, 1446, and 1458 nm were positively correlated with Ala, and 1148, 1172, 1199, 1220, 1232, 1330, 1392, 1413, 1431, and 1473 nm showed negative correlations with Ala. In the overall range, a strong positive correlation was observed at 1079, 1165, 1205, 1285, 1381, and 1446 nm, of which the peak at 1079 nm was related to the second stretching of the N-H bond in amides. The peaks at 1165, 1204, and 1285 nm were related to the second C-H stretching of carbonyl compounds. The peak at 1383 nm represents the third vibration of the C=O bond. The peak at 1446 nm represents the first O-H stretching of COOH [22–30].

It is worth noting that the regression trends of the weights of the two groups of data were similar, with 1165, 1285, and 1381 nm peaks overlapping.

3.6. Comparison of PLSR, ANN, and LS-SVM Model Effects

In order to evaluate the effect of multicombination data model, PLSR, ANN, and LS-SVM models established by feature variable data were extracted based on the weight algorithm of FS and 2D–COS, and the results are shown in Table 6. First of all, the linear PLSR model had the best effect in FS modeling, whereby $R^2_C = 0.8330$ and $R^2_P = 0.8388$, which is 9.42 and 2.74% higher than LS-SVM and ANN in the prediction set, respectively. In contrast, the modeling results of 2D–COS local data in the band with 48.9% reduction were similar to those of full band modeling, and the best modeling effect was also shown in the linear model ($R^2_C = 0.8203$, $R^2_P = 0.8190$).

Table 6. Prediction model of Ala content established by different characteristic wavelength extraction methods.

Model	Extraction Method	Variable Number	LVs	Calibration Set (n = 260)			Prediction Set (n = 87)		
				R^2_C	RMSEC	RPD_C	R^2_P	RMSEP	RPD_P
PLSR	FS	225	16	0.8330	0.1452	2.45	0.8388	0.1548	2.50
	FS–RC	26	15	0.7754	0.1684	2.11	0.8006	0.1721	2.25
	FS–CARS	36	18	0.8404	0.1419	2.51	0.8409	0.1538	2.52
	2D–COS	115	12	0.8203	0.1506	2.36	0.8190	0.1655	2.34
	2D–COS–RC	22	12	0.7536	0.1764	2.02	0.7919	0.1755	2.21
	2D–COS–CARS	36	13	0.8141	0.1531	2.33	0.8458	0.1521	2.54
LS-SVM	FS	225	-	0.7226	0.1876	1.90	0.7598	0.1907	2.03
	FS–RC	26	-	0.7323	0.1839	1.94	0.7446	0.1960	1.97
	FS–CARS	36	-	0.7278	0.1858	1.92	0.7596	0.1900	2.04
	2D–COS	115	-	0.8145	0.1600	2.23	0.7898	0.1657	2.34
	2D–COS–RC	22	-	0.7938	0.1704	2.09	0.7702	0.1747	2.22
	2D–COS–CARS	36	-	0.8212	0.1629	2.19	0.7980	0.1633	2.37
ANN	FS	225	-	0.8317	0.1461	2.44	0.8158	0.1650	2.35
	FS–RC	26	-	0.8275	0.1476	2.41	0.7785	0.1810	2.14
	FS–CARS	36	-	0.8492	0.1381	2.58	0.8312	0.1580	2.45
	2D–COS	115	-	0.8095	0.1566	2.27	0.7924	0.1767	2.19
	2D–COS–RC	22	-	0.7786	0.1672	2.13	0.7652	0.1863	2.08
	2D–COS–CARS	36	-	0.8484	0.1383	2.57	0.8341	0.1560	2.48

In the data modeling of feature wavelength extraction, the number of feature variables selected by RC was less than for CARS, but the modeling result was poor in the entire dataset. This was partly due to the negative correlation between the selected bands and Ala content. In addition, some useful information related to Ala content was lost in the process of manually selecting the absolute value of the regression coefficient to maximize, which made the model less adaptable. In variable selection modeling using the CARS algorithm, the same number of 36 characteristic variables were selected in FS and 2D–COS data, respectively, which effectively reduced data redundancy and obtained better model performance. Surprisingly, all models of CARS in the two data intervals were better than the original data and RC feature extraction data. This shows that the key bands determined by the CARS algorithm are rich in information and highly correlated with Ala content. The main reason for this is that CARS establishes a calibration model for each variable separately during operation, and calculates the weight of the regression coefficient in each variable. Different from RC algorithm, in order to avoid subjective selection, an exponential decreasing function (EDF) was used to reduce the variable space. Furthermore, adaptive reweighted sampling (ARS) was used to reduce the number of variables. Finally, the optimal variable quantum set with the minimum RMSECV was retained [35]. To put it another way, when the weighting methods were inconsistent, there were huge differences in the model effects.

Further discussing the modeling performance of CARS with the two datasets, we found that the three models based on 2D–COS local data were better than models based on FS data. The 2D–COS–CARS–PLSR model was the best (R^2_C = 0.8141, R^2_P = 0.8458). This provides the possibility that although the sensitive range of 2D–COS data compared with FS data had been reduced, the modeling effect was similar and the model accuracy had not been further improved. It was mainly used to select local intervals where data overlap and redundancy still existed. When CARS was used to eliminate useless variables, local data based on 2D–COS was effectively improved. Similarly, Fan et al. [27] used NIR–HSI to predict the TBA content of mutton, and the best prediction effect was obtained through the characteristic variable of 2D–COS local sensitive interval selected by the CARS algorithm, but the effect of 2D–COS data was slightly worse than that of FS data.

We also used RPD as the evaluation index of model performance: when RPD < 1.5, the model was invalid; when $1.5 \leq$ RPD ≤ 2.0, the model could distinguish low and high content samples; when 2 < RPD < 2.5, the model could be used as a semi-quantitative evaluation sample; when RPD $\geq$ 2.5, the model could be used for quantitative evaluation. We found that the RPDc range of Ala overall modeling was 1.90–2.58, and the RPDp range was 1.97–2.54, which indicates that the three models had good adaptability. In particular, the RPDp of linear PLSR in 2D–COS–CARS data reached 2.54, which could effectively quantitatively evaluate the content of Ala.

3.7. Visualization of Alanine Content in Beef

A visual map of Ala content was developed based on the dot product of the calculated pixel points and the weight coefficient of the optimal correction model. The optimal pretreatment MF and the optimal simplified model PLSR–2D–COS–CARS were obtained through analysis. A distribution map of the predicted Ala content of each pixel was generated by multiplying the corresponding characteristic wavelength variable and the weight coefficient. The visual variation range of Ala values (5.04–10.59 mg/100 g) is shown in Figure 7. The level of content was indicated by linear chromaticity band. The red represents high value, and blue represents low value. The Ala content in the visualization chart had obvious color difference. From Figure 7(I)–(VI), the Ala content transitions from red to yellow and finally to blue, accompanied by a decrease in content. This phenomenon might be attributed to the complex changes in compounds, including protein decomposition, lipid oxidation, and water loss. In addition, some vertical stripes appeared in the Ala visualization diagram, which might be caused by jitter noise in the line scanning of the spectrum [30]. It can be observed that the visual distribution map was evenly distributed and had clear texture, and could collect more effective spectral information. Therefore, the visualization of Ala content is of great significance for intuitively and comprehensively evaluating the dynamic changes in beef quality and nutrition.

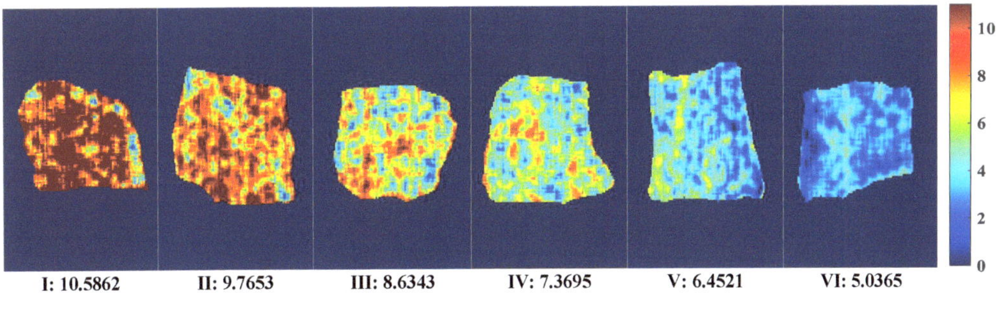

Figure 7. Visual distribution of Ala content.

4. Conclusions

Ala content could be predicted by global and local data of NIR–HSI in this study. In order to improve the spectral resolution, 2D–COS was introduced to analyze the changes in spectral characteristics, reveal the change order among various groups, and find the best local research area. The spectral data of 1000–1700 nm (225 wave bands) were used as global data, and 1136–1478 nm (115 wave bands) were used as local sensitive area data of Ala content selected by 2D–COS. On this basis, in order to improve the prediction accuracy of spectral data, the adaptive effects of the two weight algorithms in PLSR, LS-SVM, and ANN models were compared. Based on the 2D–COS–CARS–PLSR model, the optimal modeling effect was achieved (R^2_C was 0.8141, R^2_P was 0.8458, RPDp was 2.54), which was 0.83 and 3.71% higher than FS and 2D–COS data in the prediction set. The best combination prediction model was used to generate the distribution map of Ala content prediction value. The results of this study provide a possible method to predict the Ala content of beef and better explain the changes in spectral characteristics in meat product quality.

Author Contributions: F.D.: Conceptualization, Investigation, Writing Original Draft, Data Curation, Methodology. Y.B.: Writing Original Draft, Investigation, Review, and Editing. J.H.: Formal Analysis. Validation, Supervision, Review, and Editing. S.L., Y.H., and Y.L.: Validation, Methodology, Supervision. S.W.: Investigation, Writing—Review and Editing. J.C. and A.R.-G.: Project Administration. All authors have read and agreed to the published version of the manuscript.

Funding: This study was supported by the Ningxia Hui Autonomous Region fund (No. 2022AAC05022) and Key R&D Projects in Ningxia Hui Autonomous Region (No. 2019BEH03002).

Institutional Review Board Statement: Not applicable.

Informed Consent Statement: Not applicable.

Data Availability Statement: The data presented in this study are available on request from the corresponding author.

Conflicts of Interest: The authors declare no conflict of interest.

References

1. Vahmani, P.; Ponnampalam, E.N.; Kraft, J.; Mapiye, C.; Bermingham, E.N.; Watkins, P.J.; Proctor, S.D.; Dugan, M.E.R. Bioactivity and health effects of ruminant meat lipids. Invited Review. *Meat Sci.* **2020**, *165*, 108114. [CrossRef] [PubMed]
2. Bi, Y.Z.; Luo, Y.L.; Luo, R.M.; Ji, C.; Gao, S.; Bai, S.; Wang, Y.Z.; Dong, F.J.; Hu, X.L.; Guo, J.J. High freezing rate improves flavor fidelity effect of hand grab mutton after short-term frozen storage. *Front. Nutr.* **2022**, *9*, 959824. [CrossRef] [PubMed]
3. Bai, S.; You, L.Q.; Ji, C.; Zhang, T.G.; Wang, Y.R.; Geng, D.; Gao, S.; Bi, Y.Z.; Luo, R.M. Formation of volatile flavor compounds, maillard reaction products and potentially hazard substance in China stir-frying beef sao zi. *Food Res. Int.* **2022**, *159*, 111545. [CrossRef] [PubMed]
4. Alves, L.A.A.D.S.; Lorenzo, J.M.; Gonçalves, C.A.A.; Santos, B.A.D.; Heck, R.T.; Cichoski, A.J.; Campagnol, P.C.B. Impact of lysine and liquid smoke as flavor enhancers on the quality of low-fat Bologna-type sausages with 50% replacement of NaCl by KCl. *Meat Sci.* **2017**, *123*, 50–56. [CrossRef]
5. Zhang, L.; Hu, Y.Y.; Wang, Y.; Kong, B.H.; Chen, Q. Evaluation of the flavour properties of cooked chicken drumsticks as affected by sugar smoking times using an electronic nose, electronic tongue, and HS-SPME/GC-MS. *LWT-Food Sci. Technol.* **2020**, *140*, 110764. [CrossRef]
6. Chong, S.H.; Ham, S. Site-directed analysis on protein hydrophobicity. *J. Comput. Chem.* **2014**, *35*, 1364–1370. [CrossRef]
7. Yue, J.; Zhang, Y.F.; Jin, Y.F.; Deng, Y.; Zhao, Y.Y. Impact of high hydrostatic pressure on non-volatile and volatile compounds of squid muscles. *Food Chem.* **2016**, *194*, 12–19. [CrossRef]
8. Yin, M.Y.; Matsuoka, R.; Yanagisawa, T.; Xi, Y.C.; Zhang, L.; Wang, X.C. Effect of different drying methods on free amino acid and flavor nucleotides of scallop (*patinopecten yessoensis*) adductor muscle. *Food Chem.* **2022**, *396*, 133620. [CrossRef]
9. Zhang, Y.W.; Guo, X.Y.; Peng, Z.Q.; Jamali, M.A. A review of recent progress in reducing NaCl content in meat and fish products using basic amino acids. *Trends Food Sci. Technol.* **2022**, *119*, 215–226. [CrossRef]
10. Tian, Z.M.; Cui, Y.Y.; Lu, H.J.; Wang, G.; Ma, X.Y. Effect of long-term dietary probiotic *Lactobacillus reuteri* 1 or antibiotics on meat quality, muscular amino acids and fatty acids in pigs. *Meat Sci.* **2021**, *171*, 108234. [CrossRef]
11. Monteiro, S.T.; Minekawa, Y.; Kosugi, Y.; Kosugi, Y.; Akazawa, T.; Oda, T. Prediction of sweetness and amino acid content in soybean crops from hyperspectral imagery. *ISPRS J. Photogram.* **2007**, *62*, 2–12. [CrossRef]
12. Huang, H.P.; Hu, X.J.; Tian, J.P.; Jiang, X.N.; Luo, H.B.; Huang, D. Rapid detection of the reducing sugar and amino acid nitrogen contents of Daqu based on hyperspectral imaging. *J. Food Compos. Anal.* **2021**, *101*, 103970. [CrossRef]

13. Yamashita, H.; Sonobe, R.; Hirono, Y.; Morita, A.; Ikka, T. Potential of spectroscopic analyses for non-destructive estimation of tea quality-related metabolites in fresh new leaves. *Sci. Rep.* **2021**, *11*, 4169. [CrossRef]
14. Kjær, A.; Nielsen, G.; Staerke, S.; Clausen, M.R.; Edelenbos, M.; Jørgensenet, B. Prediction of Starch, Soluble Sugars and Amino Acids in Potatoes (*Solanum tuberosum* L.) Using Hyperspectral Imaging, Dielectric and LF-NMR Methodologies. *Potato Res.* **2016**, *59*, 357–374. [CrossRef]
15. Talens, P.; Mora, L.; Morsy, N.; Barbin, D.F.; ElMasry, G.; Sun, D.W. Prediction of water and protein contents and quality classification of Spanish cooked ham using NIR hyperspectral imaging. *J. Food Eng.* **2013**, *117*, 272–280. [CrossRef]
16. Wang, S.N.; Das, A.K.; Pang, J.; Liang, P. Artificial intelligence empowered multispectral vision based system for non-contact monitoring of large yellow croaker (*Larimichthys crocea*) fillets. *Foods.* **2021**, *10*, 1161. [CrossRef] [PubMed]
17. Wang, C.X.; Wang, S.L.; He, X.G.; Wu, L.G.; Li, Y.L.; Guo, J.H. Combination of spectra and texture data of hyperspectral imaging for prediction and visualization of palmitic acid and oleic acid contents in lamb meat. *Meat Sci.* **2020**, *169*, 108194. [CrossRef]
18. Ma, J.; Sun, D.W.; Pu, H.B.; Wei, Q.Y.; Wang, X.M. Protein content evaluation of processed pork meats based on a novel single shot (snapshot) hyperspectral imaging sensor. *J. Food Eng.* **2019**, *240*, 207–213. [CrossRef]
19. Dixit, Y.; Al-Sarayreh, M.; Craigie, C.R.; Reis, M.M. A global calibration model for prediction of intramuscular fat and pH in red meat using hyperspectral imaging. *Meat Sci.* **2021**, *181*, 108405. [CrossRef]
20. Ma, J.; Sun, D.W.; Pu, H.B. Model improvement for predicting moisture content (MC) in pork longissimus dorsi muscles under diverse processing conditions by hyperspectral imaging. *J. Food Eng.* **2017**, *196*, 65–72. [CrossRef]
21. Nubiato, K.E.Z.; Mazon, M.R.; Antonelo, D.S.; Calkins, C.R.; Naganathan, G.K.; Subbiah, J.; Silva, S. A bench-top hyperspectral imaging system to classify beef from Nellore cattle based on tenderness. *Infrared Phys. Technol.* **2018**, *89*, 247–254. [CrossRef]
22. Zhang, J.J.; Ma, Y.H.; Liu, G.S.; Fan, N.Y.; Li, Y.; Sun, Y.R. Rapid evaluation of texture parameters of Tan mutton using hyperspectral imaging with optimization algorithms. *Food Control* **2022**, *135*, 108815. [CrossRef]
23. Cheng, W.W.; Sørensen, K.M.; Engelsen, S.B.; Sun, D.W.; Pu, H.B. Lipid oxidation degree of pork meat during frozen storage investigated by near-infrared hyperspectral imaging: Effect of ice crystal growth and distribution. *J. Food Eng.* **2019**, *263*, 311–319. [CrossRef]
24. Cheng, W.W.; Sun, D.W.; Cheng, J.H. Pork biogenic amine index (BAI) determination based on chemometric analysis of hyperspectral imaging data. *LWT-Food Sci. Technol.* **2016**, *73*, 13–19. [CrossRef]
25. Wan, G.L.; Liu, G.S.; He, J.G.; Luo, R.M.; Cheng, L.J.; Ma, C. Feature wavelength selection and model development for rapid determination of myoglobin content in nitrite-cured mutton using hyperspectral imaging. *J. Food Eng.* **2020**, *287*, 110090. [CrossRef]
26. Cheng, W.W.; Sun, D.W.; Pu, H.B.; Wei, Q.Y. Heterospectral two-dimensional correlation analysis with near-infrared hyperspectral imaging for monitoring oxidative damage of pork myofifibrils during frozen storage. *Food Chem.* **2018**, *248*, 119–127. [CrossRef]
27. Fan, N.Y.; Liu, G.S.; Wan, G.L.; Ban, J.J.; Yuan, R.R.; Sun, Y.R.; Li, Y. A combination of near infrared hyperspectral imaging with two-dimensional correlation analysis for monitoring the content of biogenic amines in mutton. *Int. J. Food Sci. Technol.* **2021**, *56*, 3066–3075. [CrossRef]
28. Wang, Y.; Wang, C.X.; Dong, F.J.; Wang, S.L. Integrated spectral and textural features of hyperspectral imaging for prediction and visualization of stearic acid content in lamb meat. *Anal. Methods* **2021**, *13*, 4157–4168. [CrossRef]
29. Fan, N.Y.; Ma, X.; Liu, G.S.; Ban, J.J.; Yuan, R.R.; Sun, Y.R. Rapid determination of TBARS content by hyperspectral imaging for evaluating lipid oxidation in mutton. *J. Food Compos. Anal.* **2021**, *103*, 104110. [CrossRef]
30. Yu, H.D.; Qing, L.W.; Yan, D.T.; Xia, G.H.; Zhang, C.H.; Yun, Y.H.; Zhang, W.M. Hyperspectral imaging in combination with data fusion for rapid evaluation of tilapia fillet freshness. *Food Chem.* **2021**, *348*, 129129. [CrossRef]
31. Noda, I. Recent advancement in the field of two-dimensional correlation spectroscopy. *J. Mol. Struct.* **2008**, *883*, 2–26. [CrossRef]
32. Noda, I. Two-dimensional correlation analysis of spectra collected without knowing sampling order. *J. Mol. Struct.* **2018**, *1156*, 418–423. [CrossRef]
33. Dong, F.J.; Hao, J.; Luo, R.M.; Zhang, Z.F.; Wang, S.L.; Wu, K.N.; Liu, M.Q. Identification of the proximate geographical origin of wolfberries by two-Dimensional correlation spectroscopy combined with deep learning. *Comput. Electron. Agric.* **2022**, *198*, 107027. [CrossRef]
34. Yun, Y.H.; Li, H.D.; Deng, B.C.; Cao, D.S. An overview of variable selection methods in multivariate analysis of near-infrared spectra. *TrAC-Trend. Anal. Chem* **2019**, *113*, 102–115. [CrossRef]
35. Fu, J.S.; Yu, H.D.; Chen, Z.; Yun, Y.H. A review on hybrid strategy-based wavelength selection methods in analysis of near-infrared spectral data. *Infrared Phys. Technol.* **2022**, *125*, 104231. [CrossRef]
36. Cheng, L.J.; Liu, G.S.; He, J.G.; Wan, G.L.; Ma, C.; Ban, J.J.; Ma, L.M. Non-Destructive assessment of the myoglobin content of Tan sheep using hyperspectral imaging. *Meat Sci.* **2020**, *167*, 107988. [CrossRef]
37. Zhuang, Q.B.; Peng, Y.K.; Yang, D.Y.; Wang, Y.L.; Zhao, R.H.; Chao, K.L.; Guo, Q.H. Detection of frozen pork freshness by fluorescence hyperspectral image. *J. Food Eng.* **2022**, *316*, 110840. [CrossRef]

biosensors

MDPI

Article

Optimization of Sensors Data Transmission Paths for Pest Monitoring Based on Intelligent Algorithms

Yuyang Lian [1,2,†], Aqiang Wang [1,†], Sihua Peng [1,2,†], Jingjing Jia [3,†], Liang Zong [4,†], Xiaofeng Yang [1,†], Jinlei Li [1], Rongjiao Zheng [1], Shuyan Yang [1], Jianjun Liao [1] and Shihao Zhou [1,*]

1 Sanya Nanfan Research Institute of Hainan University, Sanya 572025, China
2 Key Laboratory of Germplasm Resources Biology of Tropical Special Ornamental Plants of Hainan Province, College of Forestry, Hainan University, Haikou 570228, China
3 Hainan Key Laboratory for Control of Plant Diseases and Insect Pests, Haikou 571199, China
4 College of Information Engineering, Shaoyang University, Shaoyang 422000, China
* Correspondence: zsh88200939@126.com
† These authors contributed equally to this work.

Abstract: The harm of agricultural pests presents a remarkable effect on the quality and safety of edible farm products and the monitoring and identification of agricultural pests based on the Internet of Things (IoT) produce a large amount of data to be transmitted. To achieve efficient and real-time transmission of the sensors' data for pest monitoring, this paper selects 235 geographic coordinates of agricultural pest monitoring points and uses genetic algorithm (GA), particle swarm optimization (PSO), and simulated annealing (SA) to optimize the data transmission paths of sensors. The three intelligent algorithms are simulated by MATLAB software. The results show that the optimized path based on PSO can make the shortest time used for transmitting data, and its corresponding minimum time is 4.868012 s. This study can provide a reference for improving the transmission efficiency of agricultural pest monitoring data, provide a guarantee for developing real-time and effective pest control strategies, and further reduce the threat of pest damage to the safety of farm products.

Keywords: pest monitoring; sensors; data transmission; genetic algorithms; particle swarm optimization; simulated annealing

Citation: Lian, Y.; Wang, A.; Peng, S.; Jia, J.; Zong, L.; Yang, X.; Li, J.; Zheng, R.; Yang, S.; Liao, J.; et al. Optimization of Sensors Data Transmission Paths for Pest Monitoring Based on Intelligent Algorithms. *Biosensors* **2022**, *12*, 948. https://doi.org/10.3390/bios12110948

Received: 22 September 2022
Accepted: 26 October 2022
Published: 1 November 2022

1. Introduction

The use of large amounts of fossil fuels has accelerated global warming since the 20th century, and atmospheric temperatures and sea levels are rising at an unprecedented rate [1]. It is expected that the global average temperature will rise by 1.5–4.5 °C by the end of this century [2]. Insects are typically poikilotherm, and temperature is the most important environmental factor affecting insect population dynamics. Climate warming present a series of influences on pests, such as expanding the range of fitness zones [3], increasing the survival rate of overwintering, increasing the number of generations [4], and increasing the risk of spreading crop diseases. These effects may cause sudden local outbreaks or migrations of pest populations, reduce crop yields, affect normal crop growth and development, and even cause massive crop mortality and severe crop failure, thus presenting a serious threat to the quality and safety of agricultural products. Once crop yields are severely affected, it can cause local crop supply and demand to be relatively tight, making the price of food products remarkable high and hindering the steady improvement of a region's agricultural economy. This means that the agricultural pests cause huge economic losses as well as presenting a remarkable threat to food safety [5,6]. Only through effective monitoring methods to achieve early warning of pest occurrence can we reduce agricultural losses and ensure the safety of edible farm products.

Traditionally, the acquisition of pest information is mainly performed by manual field survey and identification statistics, and the monitoring of pest situations is based on the manual survey, which is a method with high workload, low efficiency, poor reliability, and low accuracy [7,8]. Agricultural pest monitoring and identification using the Internet of Things (IoT) has become an important component of the agricultural pest control category with the advent of Industry 4.0. Combining agriculture with modern electronic information industries such as communications and sensors has greatly improved the efficiency and accuracy of pest monitoring and identification. Gassoumi et al. designed a computer vision-based insect classification and identification system for cotton fields, using an artificial neural network approach to classify 12 insect species in cotton fields, the recognition rate of 11 insect species reached more than 90%, and the recognition rate of only one insect was 72% [9]. Based on the existing light trapping technology, an automatic pest collection device that can be used for cruciferous major insect pest detection was studied and designed. Using this device, images of cruciferous major pests can be collected, flipped, and vibrated according to the real-time situation of pest images, which effectively improves the quality of image collection and recognition accuracy, and the recognition error is less than 5% [10]. Based on a wireless network image sensors system, a pest remote automatic monitoring and early warning system was designed, which uses the technique of background differencing to achieve pest counting and sends an early warning message when the pest density is greater than a threshold value [11]. Pest monitoring and identification technology are continuously optimized and the accuracy is continuously improved, but multiple monitoring points are often set up in the process of agricultural pest monitoring in the field. The network will be blocked when the number of monitoring points increases or during the peak of pest occurrence, which will delay or collapse the data transmission, thus limiting the real-time monitoring of agricultural pests. As a result, the information on the species and quantity of pests cannot be obtained in time, and effective pest control strategies cannot be formulated. Therefore, it is urgent to increase research on the optimization of sensors' data transmission paths for pest monitoring to achieve efficient and real-time transmission of multi-point data in the process of pest monitoring.

Genetic algorithm (GA), swarm optimization (PSO), and simulated annealing (SA) are all bionic intelligence algorithms which are widely used in industry, agriculture, and medicine to find optimal solutions [11]. GA, also known as the evolutionary algorithm (EA), is a heuristic algorithm based on the process of biological evolution. Its main feature is the selection of the solution to the problem based on the "survival of the fittest" or "competition" method of biological evolution [12]. GA is based on all individuals in a population and uses randomization techniques to guide an efficient search of coded parameter space. Among them, selection, crossover, and mutation constitute the genetic operations of GA; five elements, namely, parameter encoding, initial population setting, design of fitness function, design of genetic operations, and setting of control parameters, form the core of GA [13]. GA has been widely used in the fields of combinatorial optimization, machine learning, signal processing, adaptive control, and artificial life, and it is one of the key techniques in modern relevant intelligent computing [14]. Zhang tested the performance of the improved GA on the transmission path of wireless Mesh network through simulation to optimize the transmission path of marine data and improve the efficiency of marine data transmission [15]. PSO, also known as the birds' flock foraging algorithm, is a new EA developed by Kennedy and Eberhart et al. in recent years. The PSO starts from a random solution, finds the optimal solution by iteration, evaluates the quality of the solution by fitness, and follows the current searched optimal value to find the global optimal. This algorithm has attracted the attention of academics for its easy implementation, high accuracy, and fast convergence, and has demonstrated its superiority in solving practical problems [16]. Ma et al. applied the PSO to the medical field, and the research based on the PSO will use a multi-PSO for path planning to effectively reduce the path search range and improve the path search speed [17]. SA is a general probabilistic heuristic algorithm for combinatorial optimization problems to find a global optimal solution and an approximate optimal solution in a large global search space [18]. The SA is often used for searching in discrete spaces; it will be more effective than the exhaustive method for some problems because it aims to find a good solution in an acceptable time, rather than the best solution [19]. Based on the SA, Liu et al. proposed an optimal design method for the heat extraction section of the energy tunnel, which effectively improved the heat transfer efficiency and reduced the construction cost [20]. The sensors' data transmission paths for pest monitoring are optimized based on GA, PSO, and SA in our study (Figure 1), and the data transmission speed is compared and analyzed after the optimized paths, which improves the transmission efficiency of agricultural pest monitoring data, provides a guarantee for the formulation of effective agricultural pest control strategies, and reduces the threat of pest damage to the quality and safety of farm products.

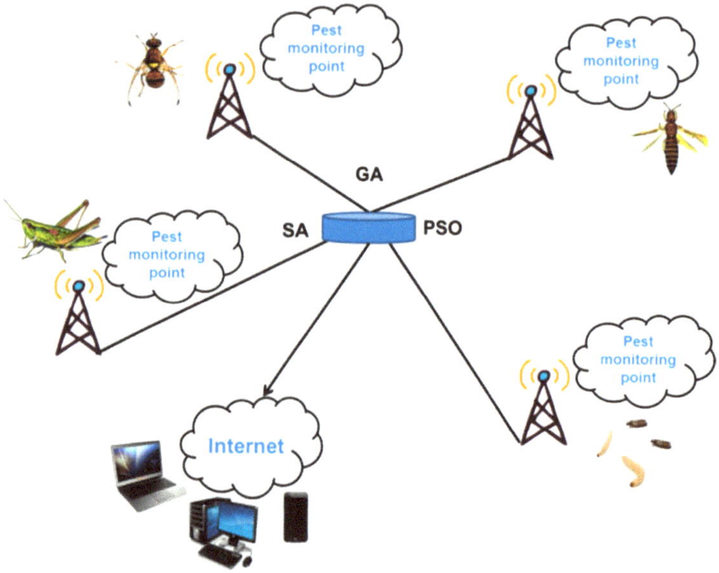

Figure 1. The sensors' data transmission paths for multiple pest monitoring points are optimized based on genetic algorithm, particle swarm optimization, and simulated annealing.

2. Methods

In this paper, 235 geographic location coordinates of agricultural pest monitoring points are selected, and the sensors' data transmission paths for pest monitoring points are optimized and analyzed based on GA, PSO, and SA, and the optimal transmission path and transmission time are determined.

2.1. Genetic Algorithm

The GA is used to analyze the prediction of their weights.

Step1. Encode the data:

$$G_k = [V_1, V_2, \ldots, V_t, \ldots, V_T] = [R_1, R_2, \ldots, R_i, \ldots, R_n]^T = \begin{bmatrix} p_{11} & p_{12} & \cdots & p_{1t} & \cdots & p_{1T} \\ p_{21} & p_{22} & \cdots & p_{2t} & \cdots & p_{2T} \\ \vdots & \vdots & \cdots & \vdots & \cdots & \vdots \\ p_{i1} & p_{i2} & \cdots & p_{it} & \cdots & p_{iT} \\ \vdots & \vdots & \cdots & \vdots & \cdots & \vdots \\ p_{n1} & p_{n2} & \cdots & p_{nt} & \cdots & p_{nT} \end{bmatrix} \tag{1}$$

$$u_{it} = \begin{cases} 0, & p_{it} = 0 \\ 1, & \text{other} \end{cases} \tag{2}$$

Step2. Coding during initialization of genetic populations:

$$\mathrm{R} = [\mathrm{r}_1, r_2, \ldots, r_i, \ldots, r_n]\ i = 1, 2, \ldots, n \tag{3}$$

$$per = [per_1, per_2, \ldots, per_i, \ldots, per_n] \tag{4}$$

$$\begin{aligned} per_i &= \frac{r_i}{\sum\limits_{i=1}^{n} r_i} \quad i = 1, 2, \ldots, n \\ V_t &= [p_{1t}, p_{2t}, \ldots, p_{it}, \ldots, p_{nt}]^T \end{aligned} \tag{5}$$

Continue to optimize and simulate it.
Step3. Adjustment for each individual adaptation:

$$p_{it}{}^1 = \begin{cases} p_{i\max}, & p_{it} > p_{i\max} \\ p_{it}, & p_{i\min} \leq p_{it} \leq p_{i\max} \\ p_{i\min}, & \lambda p_{i\min} \leq p_{it} \leq p_{i\max} \\ 0, & \text{other} \end{cases} \tag{6}$$

Step4. The function takes the following form:

$$f_{itness}(G_k) = \frac{A}{(F + \sum\limits_{i}^{n} m\delta S_i)} \tag{7}$$

Step5. The selection probability is calculated by:

$$P(x_i) = \frac{f_{itness}(x_i)}{\sum\limits_{j=1}^{n} f_{itness}(x_j)} \tag{8}$$

$$q_i = \sum_{j=1}^{i} P(x_j) \tag{9}$$

Step6. Cross mutation and offspring combination were carried out:

$$\begin{aligned} C_1 &= G^{C1} = [V_1{}^{C1}, V_2{}^{C1}, \ldots, V_t{}^{C1}, \ldots, V_T{}^{C1}] \\ C2 &= G^{C2} = [V_1{}^{C2}, V_2{}^{C2}, \ldots, V_t{}^{C2}, \ldots, V_T{}^{C2}] \end{aligned} \tag{10}$$

The specific operation process is:

$$\begin{aligned} D_1 &= [V_1{}^{C1}, V_2{}^{C1}, \ldots, V_{j-1}{}^{C1}, (1-\alpha)V_j{}^{C1} + \alpha V_j{}^{C2}, V_{j+1}{}^{C1}, \ldots, V_T{}^{C1}] \\ D_2 &= [V_1{}^{C2}, V_2{}^{C2}, \ldots, V_{j-1}{}^{C2}, \alpha V_j{}^{C1} + (1-\alpha)V_j{}^{C2}, V_{j+1}{}^{C2}, \ldots, V_T{}^{C2}] \end{aligned} \tag{11}$$

$$\begin{aligned} O_1 &= \max = \{f_{itness}(C_1), f_{itness}(D_1)\} \\ O_2 &= \max = \{f_{itness}(C_2), f_{itness}(D_2)\} \end{aligned} \tag{12}$$

Step7. Individual crossover through Step 6:

$$\begin{aligned} E_1 &= [V_1{}^{O1}, V_2{}^{O1}, \ldots, V_t{}^{O1}, \ldots, V_T{}^{O1}] \\ E_2 &= [V_1{}^{O2}, V_2{}^{O2}, \ldots, V_t{}^{O2}, \ldots, V_T{}^{O2}] \end{aligned} \tag{13}$$

$$\eta, (\eta = \begin{cases} 1, & \text{Individuals mutated} \\ 0, & \text{Individuals did not mutate} \end{cases})$$

$$\begin{aligned} H_1 &= [V_1{}^{O1}, V_2{}^{O1}, \ldots, \eta\beta V_\gamma{}^{O1} + (1-\eta)V_\gamma{}^{O1}, V_{\gamma+1}{}^{O1}, \ldots, V_T{}^{O1}] \\ H_2 &= [V_1{}^{O2}, V_2{}^{O2}, \ldots, \eta\beta V_\gamma{}^{O2} + (1-\eta)V_\gamma{}^{O2}, V_{\gamma+1}{}^{O2}, \ldots, V_T{}^{O2}] \end{aligned} \tag{14}$$

$$\begin{aligned} I_1 &= \max = \{f_{itness}(E_1), f_{itness}(H_1)\} \\ I_2 &= \max = \{f_{itness}(E_2), f_{itness}(H_2)\} \end{aligned} \tag{15}$$

2.2. Particle Swarm Optimization

In the PSO algorithm, the particle population is searched in an n-dimensional space, where the position of each particle X_i represents a solution to the problem, and the particle searches for a solution by continuously updating its position.

The position of the *i*-th particle at moment *t* is denoted by $X_{i.t} = [x_{i.t.1}, x_{i.t.2}, \cdots, x_{i.t.n}]$.

The velocity of the first particle at moment *t* is denoted by $V_{i.t} = [v_{i.t.1}, v_{i.t.2}, \cdots, v_{i.t.n}]$.

The first particle remembers at each moment the best position it searched for before moment *t*, denoted as P_t, and the best position searched for by the whole particle population, denoted as P_g.

The velocity V_{it} of each particle at moment *t* is influenced by its own optimal position and the population optimal position, and the velocity and position of each particle are updated according to the following equation:

$$V_{i,t} = \omega V_{i,t-1} + c_1 r_1 (P_i - X_i) + c_2 r_2 (P_g - X_i)$$
$$V_{i,t} = V_{i,t-1} + V_{i,t}$$

where ω is called the power inertia factor, c_1, c_2 is a positive acceleration constant, and r_1, r_2 is a random number uniformly distributed between 0 and 1. The corresponding algorithm flowchart of the PSO is shown in Figure 2

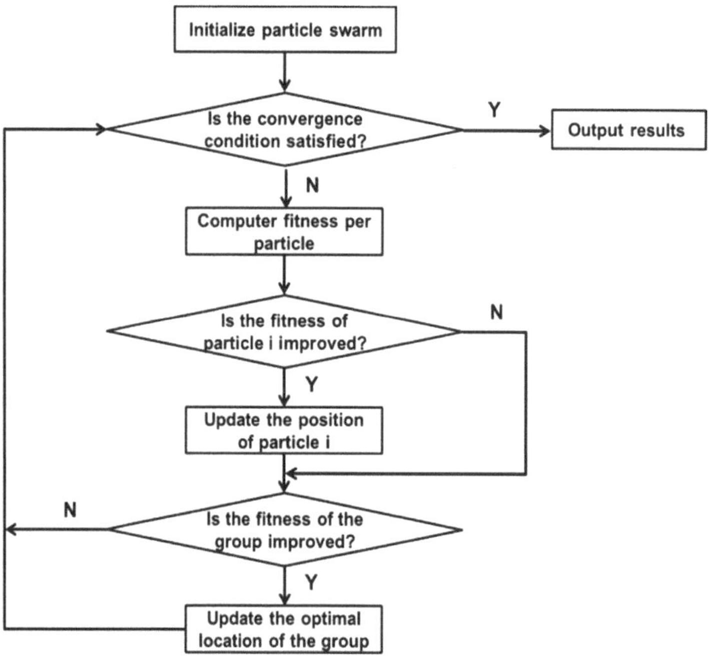

Figure 2. Description of particle swarm optimization.

2.3. Simulated Annealing

The optimization of sensors' data transmission paths for pest monitoring based on SA first determines the initial temperature, randomly selects an initial state and examines the objective function value of that state, attaches a small perturbation to the current state and calculates the objective function value of the new state, and accepts the better point with probability 1 and the worse point with some probability Pr as the current point until the system cools down. The SA is able to converge to the global optimum with probability 1 under the condition that the initial temperature is high enough and the temperature decreases slowly enough. The SA process of IoT monitoring data transmission is completed, and the regulation of monitoring network path optimization is finally realized. The corresponding algorithm flow chart is shown in Figure 3.

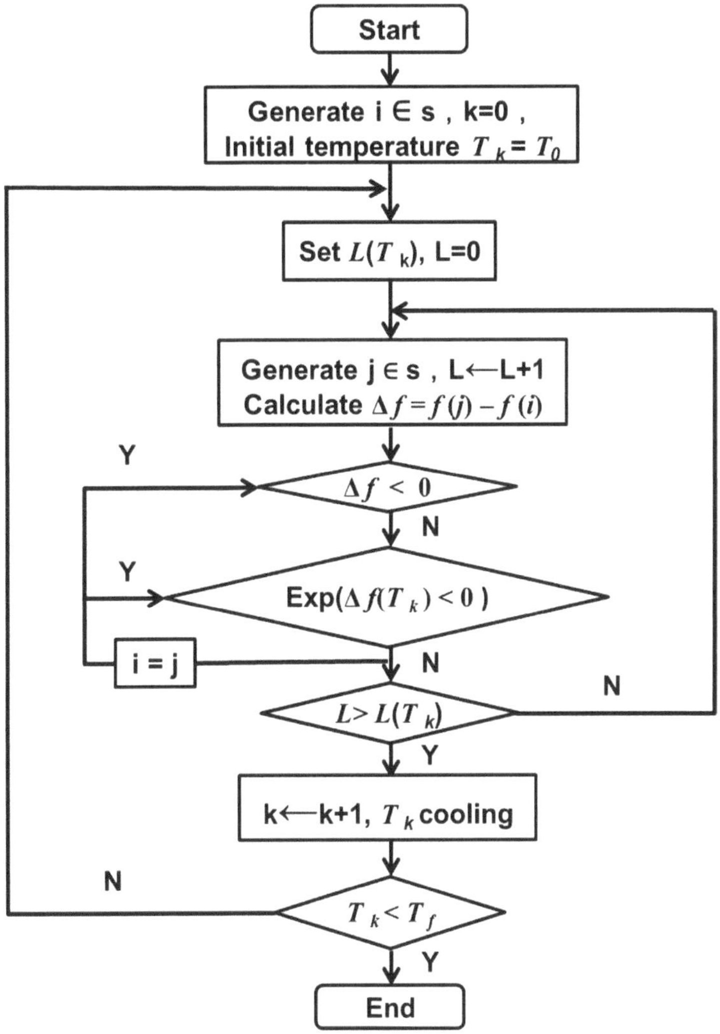

Figure 3. Description of simulated annealing.

3. Result

3.1. Genetic Algorithm

The data of 235 pest monitoring points were imported into MATLAB software (MathWorks, MA, USA), and the GA was used for optimization analysis, and the results are shown in Figures 4 and 5. In order to test whether the GA-based pest monitoring data reached the global minimum, the total number of iterations was set to 1000. The value of the objective function gradually decreased, and the number of transients of its convergence trajectory increases with the increase in the number of iterations, resulting in many local minima, which reach the global minimum at about 900 iterations.

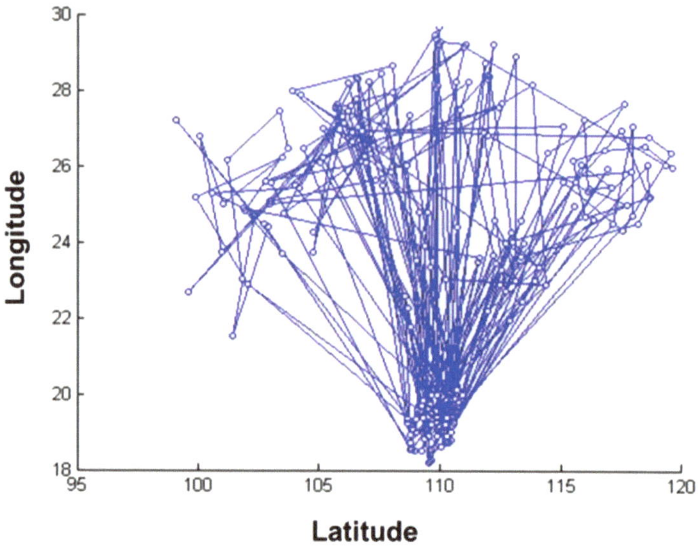

Figure 4. Running results of the genetic algorithm.

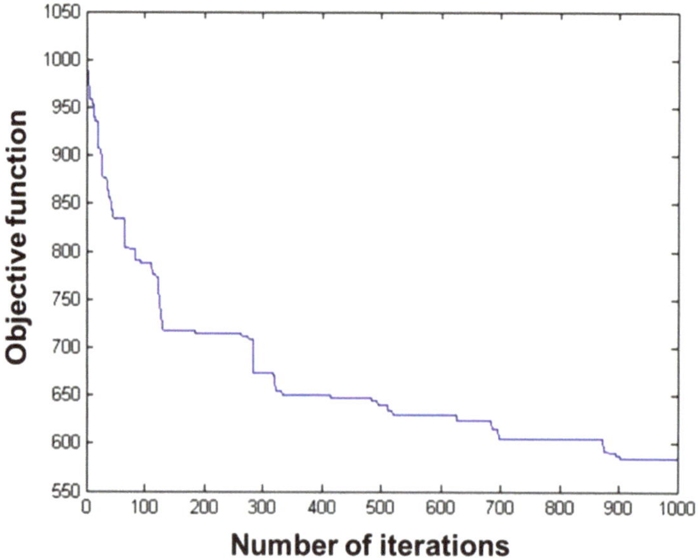

Figure 5. Convergence trajectory of the genetic algorithm.

3.2. Particle Swarm Optimization

The data of 235 pest monitoring points were imported into MATLAB software, and the PSO was used for optimization analysis, and the results are shown in Figures 6 and 7. The total number of iterations was set to 14, and the value of the objective function decreased with the increase in the number of iterations, and its convergence trajectory started to converge rapidly at the number of iterations of four, and reached the global minimum at about five iterations.

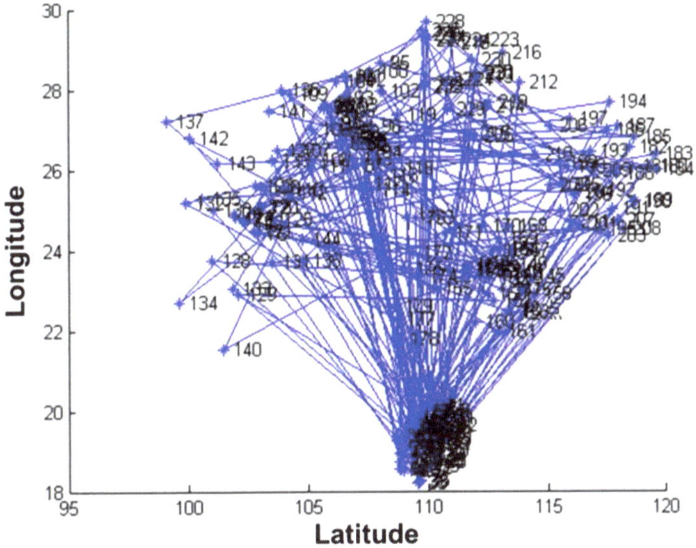

Figure 6. Running results of particle swarm optimization.

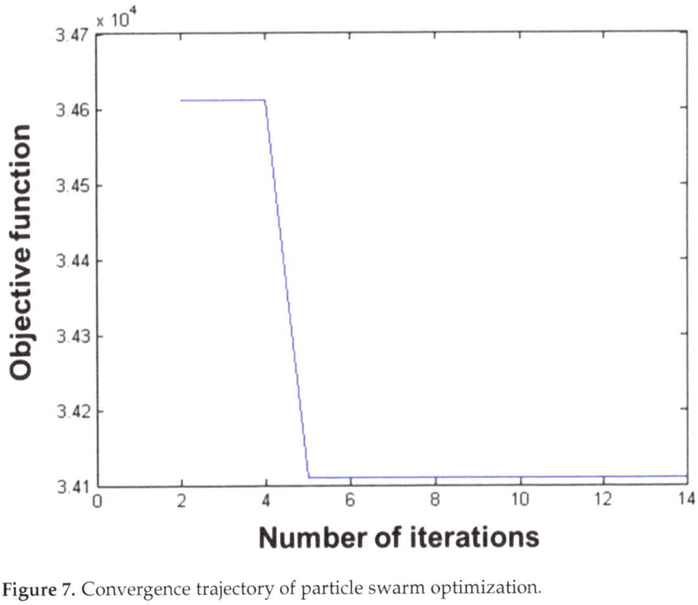

Figure 7. Convergence trajectory of particle swarm optimization.

3.3. Simulated Annealing

The data of 235 pest monitoring points were imported into MATLAB software, and the SA was used for optimization analysis, and the results are shown in Figures 8 and 9. The total number of iterations was set to 900, and the value of the objective function decreased slowly with the increase in the number of iterations, and its convergence trajectory was smoother than that of the GA, and it reached the minimum value at about 800 iterations.

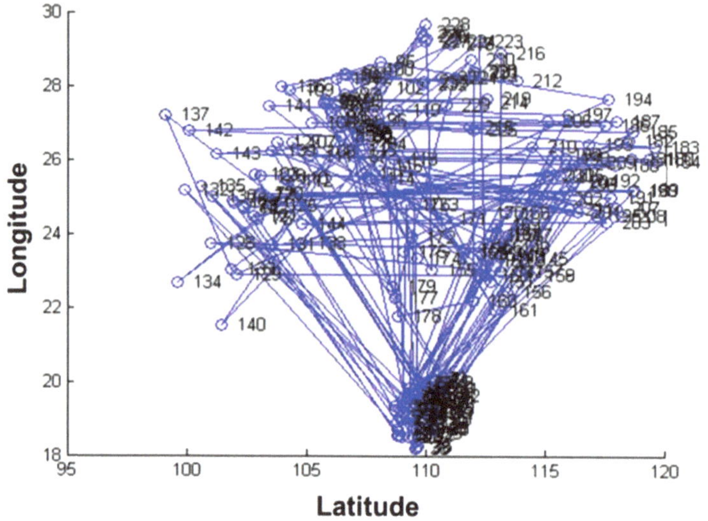

Figure 8. Running results of simulated annealing.

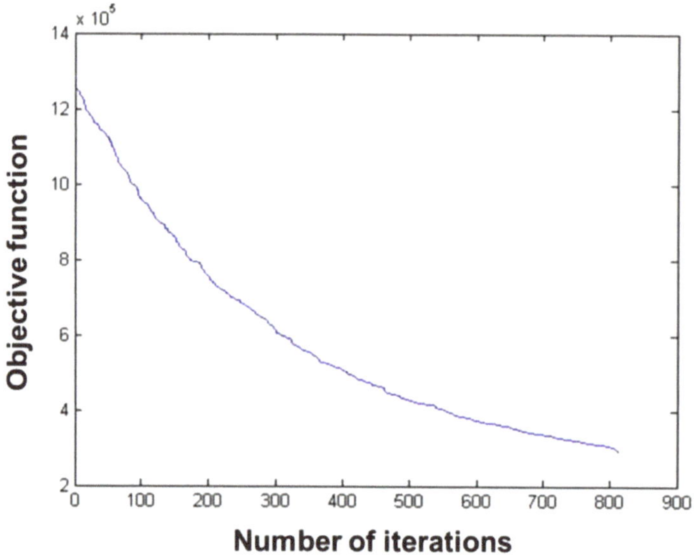

Figure 9. Convergence trajectory of simulated annealing.

The results showed that the transmission completion time of 235 pest monitoring points was 262.738048, 4.868012, and 17.842523 s, respectively, after optimizing the sensors' data transmission path based on GA, PSO, and SA (Table 1). The transmission speed after optimization based on PSO was much higher than that of GA and SA. The objective function values of all three algorithms decrease with the increase in the number of iterations, the curves gradually converge, and the objective function values are close to the optimal. The convergence trajectory of the PSO has reached the optimal value at about five iterations, and the GA and SA reach the optimal value at about 900 and 800 iterations, respectively. The convergence speed of the PSO-based algorithm is higher than that of the GA and the SA, and the convergence trajectory of the GA has more transients. This show that the performance of the PSO algorithm is better than that of the GA and the SA in the optimization of sensors' data transmission path for pest monitoring based on three algorithms.

Table 1. The elapsed time of three algorithms to complete the optimal path.

Types of Algorithms	Elapsed Time (s)
Genetic Algorithm	262.738048
Particle Swarm Optimization	4.868012
Simulated Annealing	17.842523

4. Discussion

Internet technology is a new distributed sensors network and information transmission network integrating various sensing technologies, modern communication technologies, artificial intelligence, and automatic control. It has been widely used in insect monitoring and early warning, as well as agricultural product traceability, etc. [21,22]. Agricultural pests are characterized by a wide range of species, large numbers, wide distribution, and serious damage, which makes the monitoring and integrated prevention and control of agricultural pests particularly important. In the process of monitoring agricultural pests in the field, considering the mobility and random rows of agricultural pests, a large number of monitoring points need to be set up in the monitoring area, which facilitates the construction of a sensors network for pest monitoring and maximizes the collection of pest information [23]. However, the interconnection of multiple monitoring points means that the sensors' data transmission from monitoring nodes to aggregation nodes may have a large amount of redundant data, making the monitoring network congested or even collapsing, resulting in failure to receive timely pest information. A total of 235 pest monitoring points were selected and the sensors' data transmission paths were optimized based on three bionic intelligent algorithms (GA, PSO, and SA) in our study to improve the sensors' data transmission efficiency for pest monitoring. The data transmission completion time was 262.738048 s after optimization using GA, 4.868012 s for PSO, and 17.842523 s for SA. The transmission time after optimization based on PSO was much lower than that of GA and SA, which shows that the PSO presents the best effect on the optimization of sensors' data transmission path. Zhu et al. proposed a target recognition method based on PSO for data transmission path in wireless sensor networks, with an average delay of 1.8 μs. The transmission energy consumption and delay are reduced [24].

The GA is a computational model of the biological evolutionary process that simulates the natural selection and genetics mechanism of Darwinian biological evolution, the PSO is an intelligent algorithm designed by simulating the predatory behavior of a flock of birds, and the SA is derived from the principle of solid annealing. All three algorithms are heuristic algorithms, which are mathematical simulations of natural processes, and all are global optimization-seeking algorithms [25]. The GA has a strong global search capability and can jump out of the local optimum well, but it has a slow convergence, long running time, and insufficient local search capability; the PSO is simple to operate, and does not require complex behaviors such as population selection, crossover, and mutation, it runs fast but has the risk of falling into the local optimum solution; the SA has a strong local search capability, short running time, and poor global search capability [26,27]. Premalatha et al. proposed a hybrid algorithm PSO-GA based on GA, which applied the variation of GA to PSO and avoided the PSO to fall into the local optimal solution [28]. Da et al. proposed an improved PSO algorithm based on the SA technique, and the results showed that the SAPSO-based artificial neural network presents a better ability to escape from local optimum and is more effective than the traditional PSO-based artificial neural network [29]. A hybrid algorithm using a combination of GA and SA is proposed to solve the shortest path optimization problem of undirected networks, which prevents the occurrence of premature maturation, ensures the diversity of the population, and can effectively prevent the occurrence of falling into the local optimal search situation [30]. The three algorithms are pairwise combined into the hybrid algorithm, which can effectively complement the prematurity of GA, and the disadvantages of convergence speed decrease and similarity increase due to the random search algorithm. The sensors' data transmission paths were optimized based on each of the three algorithms and compared in our study, and if the hybrid algorithm can be used in combination with the agricultural pest monitoring data transmission, the sensors' data transmission path is further optimized.

In the process of agricultural pest monitoring, the performance of sensor nodes plays a decisive role in data transmission [31]. The speed of sensors' data transmission for pest monitoring is guaranteed, and the data can be transmitted in real-time to develop pest control strategies in time after using PSO to optimize the path, but the sensor nodes are powered by batteries, which have limited energy and are extremely inconvenient to charge [32]. The lifetime of sensor nodes depends on the battery life, and excessive energy loss can cause the premature end of the agricultural pest monitoring network [33]. Therefore, while considering the transmission speed, the energy required for sensors' data transmission also needs to be considered. Yang proposed a high-energy data transmission optimization algorithm-OTCEE based on tree-cluster topology, which reduces the energy loss of sensor nodes and delays the death of nodes [34]. Abhishek et al. developed an energy-saving scheme based on PSO optimization and GA optimization techniques, which solves the sensor node energy consumption fast [35]. Therefore, the efficiency of sensors' data transmission for pest monitoring will be greatly improved if an algorithm can be designed to maximize the transmission speed and minimize the energy consumption.

5. Conclusions

In conclusion, this study uses three intelligent algorithms (GA, PSO, and SA) to optimize the data transmission paths of the sensors, and the optimization results are compared and analyzed. The results show that the optimized path based on PSO presents the shortest time used for transmitting data, and its minimum time is 4.868012 s. This study can provide a reference for improving the sensors' data transmission efficiency for agricultural pest monitoring, and guarantee the development of real-time and effective pest control programs.

Author Contributions: A.W., Y.L., S.P., J.J., X.Y., J.L. (Jinlei Li), L.Z. and S.Z. participated in the study design and analysis of the manuscript. J.L. (Jianjun Liao), R.Z. and S.Y. participated in the study design and helped to draft the manuscript. S.Z. revised and processed the manuscript and provided supervision and financial support. All authors have read and agreed to the published version of the manuscript.

Funding: This study was supported by Hainan Major Science and Technology Project (ZDKJ201901), Hainan Province Science and Technology Special Fund (ZDYF2022XDNY163), Open project of Hainan Key Laboratory for Control of Plant Diseases and Insect Pests (2022), Hunan Provincial Natural Science Foundation (2022JJ50245), Project of the Administrative Bureau of Sanya Yazhou Bay Science and Technology City.

Institutional Review Board Statement: This is an observational study. The Insects Research Ethics Committee has confirmed that no ethical approval is required.

Data Availability Statement: The datasets generated during and/or analyzed during the current study are available from the corresponding author on reasonable request.

Conflicts of Interest: The authors declare no conflict of interest.

References

1. Shrestha, S. Effects of climate change in agricultural insect pest. *Acta Sci. Agric.* **2019**, *3*, 74–80. [CrossRef]
2. Easterling, D.R.; Meehl, G.A.; Parmesan, C.; Changnon, S.A.; Karl, T.R.; Mearns, L.O. Climate extremes: Observations, modeling, and impacts. *Science* **2000**, *289*, 2068–2074. [CrossRef] [PubMed]
3. Menéndez, R. How are insects responding to global warming? *Tijdschr. Entomol.* **2007**, *150*, 355.
4. Yamamura, K.; Kiritani, K. A simple method to estimate the potential increase in the number of generations under global warming in temperate zones. *Appl. Entomol. Zool.* **1998**, *33*, 289–298. [CrossRef]
5. Sexton, S.E.; Lei, Z.; Zilberman, D. The economics of pesticides and pest control. *Environ. Res. Econ.* **2007**, *1*, 271–326. [CrossRef]

6. Lima, M.C.F.; de-Almeida-Leandro, M.E.D.; Valero, C.; Coronel, L.C.P.; Bazzo, C.O.G. Automatic detection and monitoring of insect pests—A review. *Agriculture* **2020**, *10*, 161. [CrossRef]
7. Yue, J.; Lei, T.; Li, C.; Zhu, J. The application of unmanned aerial vehicle remote sensing in quickly monitoring crop pests. *Intell. Autom. Soft Comput.* **2012**, *18*, 1043–1052. [CrossRef]
8. Preti, M.; Verheggen, F.; Angeli, S. Insect pest monitoring with camera-equipped traps: Strengths and limitations. *J. Pest. Sci.* **2021**, *94*, 203–217. [CrossRef]
9. Gassoumi, H.; Prasad, N.R.; Ellington, J.J. Neural network-based approach for insect classification in cotton ecosystems. In *International Conference on Intelligent Technologies*; InTech: Bangkok, Thailand, 2000.
10. Liu, Y.X. Study on Automatic Collection Device of Main Pests in Cruciferous Vegetables. Master's Thesis, Zhejiang Agriculture and Forestry University, Zhejiang, China, 2020.
11. Coleman, C.M.; Rothwell, E.J.; Ross, J.E. Investigation of simulated annealing, ant-colony optimization, and genetic algorithms for self-structuring antennas. *IEEE Trans. Antennas Propag.* **2004**, *52*, 1007–1014. [CrossRef]
12. Katoch, S.; Chauhan, S.S.; Kumar, V. A review on genetic algorithm: Past, present, and future. *Multimed. Tools. Appl.* **2021**, *80*, 8091–8126. [CrossRef]
13. Whitley, D. A genetic algorithm tutorial. *Stat. Comput.* **1994**, *4*, 65–85. [CrossRef]
14. Haldurai, L.; Madhubala, T.; Rajalakshmi, R. A study on genetic algorithm and its applications. *Int. J. comput. Sci. Eng.* **2016**, *4*, 139.
15. Zhang, Q. Study on Ocean Data Transmission Path Optimization and Acquisition Method in Narrowband Network. Master's Thesis, Jimei University, Fujian, China, 2020.
16. Poli, R.; Kennedy, J.; Blackwell, T. Particle swarm optimization. *Swarm Intell.* **2007**, *1*, 33–57. [CrossRef]
17. Ma, T.; Yang, Q.; Li, Z.L. Orthodontic path planning based on improved polyparticle swarm optimization. *J. Graph.* **2021**, *42*, 615–622.
18. Rutenbar, R.A. Simulated annealing algorithms: An overview. *IEEE Circuits Devices Mag.* **1989**, *5*, 19–26. [CrossRef]
19. Dowsland, K.A.; Thompson, J. Simulated annealing. *Handb. Nat. Comput.* **2012**, 1623–1655.
20. Liu, J.X.; Han, C.J.; Cai, G.Q. Optimization design of heat harvesting section of energy tunnel based on simulated annealing method. *J. Shenzhen Univ.* **2022**, *39*, 3–12. [CrossRef]
21. Kwon, S.W.; Kim, J.Y.; Cho, H.S.M.Y.; Kim, K.J. Development of wireless vibration sensor using MEMS for tunnel construction and maintenance. *Tunn. Undergr. Space Technol.* **2006**, *21*, 318. [CrossRef]
22. Huang, Q.; Tang, B.; Deng, L. Development of high synchronous acquisition accuracy wireless sensor network for machine vibration monitoring. *Measurement* **2015**, *66*, 35–44. [CrossRef]
23. Feng, H.Q.; Yao, Q. Automatic identification and monitoring technology of agricultural pests. *Plant Prot.* **2018**, *44*, 127–133.
24. Zhu, Y.Q.; Tian, E.L. Target recognition method for data transmission path in wireless sensor networks. *J. Shenyang Univ. Technol.* **2021**, *43*, 307–310.
25. Nazari-Heris, M.; Mohammadi-Ivatloo, B. Application of heuristic algorithms to optimal PMU placement in electric power systems: An updated review. *Renew. Sust. Energ. Rev.* **2015**, *50*, 214–228. [CrossRef]
26. Guo, L.; Zhao, S.; Shen, S.; Jiang, C. Task scheduling optimization in cloud computing based on heuristic algorithm. *J. Netw.* **2012**, *7*, 547. [CrossRef]
27. Zhang, Y.; Wang, S.; Ji, G. A comprehensive survey on particle swarm optimization algorithm and its applications. *Probl. Eng.* **2015**, *2015*, 931256. [CrossRef]
28. Premalatha, K.; Natarajan, A.M. Hybrid PSO and GA for global maximization. *Int. J. Open Problems Compt.* **2009**, *2*, 597–608.
29. Da, Y.; Xiurun, G. An improved PSO-based ANN with simulated annealing technique. *Neurocomputing* **2005**, *63*, 527–533. [CrossRef]
30. Wen, X.L. Shortest path optimization algorithm based on hybrid algorithm. *J. Tianjin Univ. Technol.* **2009**, *25*, 37–40.
31. Rustia, D.J.A.; Lin, C.E.; Chung, J.Y.; Zhuang, Y.J.; Hsu, J.C.; Lin, T.T. Application of an image and environmental sensor network for automated greenhouse insect pest monitoring. *J. Asia Pac. Entomol.* **2020**, *23*, 17–28. [CrossRef]
32. Anastasi, G.; Conti, M.; Di, F.M.; Passarella, A. Energy conservation in wireless sensor networks: A survey. *Ad Hoc Netw.* **2009**, *7*, 537–568. [CrossRef]
33. Sun, Y.; Dong, W.; Chen, Y. An improved routing algorithm based on ant colony optimization in wireless sensor networks. *IEEE Commun. Lett.* **2017**, *21*, 1317–1320. [CrossRef]
34. Yang, Y.Q. An efficient data transmission optimization algorithm in wireless sensor networks. *South. Farm Mach.* **2021**, *52*, 112–114.
35. Agnihotri, A.; Gupta, I.K. A hybrid PSO-GA algorithm for routing in wireless sensor network. In Proceedings of the 2018 4th International Conference on Recent Advances in Information Technology (RAIT), Dhanbad, India, 15–17 March 2018; pp. 1–6.

biosensors

MDPI

Article

An Apple Fungal Infection Detection Model Based on BPNN Optimized by Sparrow Search Algorithm

Changtong Zhao [1], Jie Ma [1], Wenshen Jia [1,2,*], Huihua Wang [3], Hui Tian [1], Jihua Wang [2] and Wei Zhou [4]

[1] Mechanical Electrical Engineering School, Beijing Information Science and Technology University, Beijing 100192, China
[2] Institute of Quality Standard and Testing Technology, Beijing Academy of Agriculture and Forestry Sciences, Beijing 100097, China
[3] Department of Food and Bioengineering, Beijing Vocational College of Agriculture, Beijing 102206, China
[4] Hebei Food Safety Key Laboratory, Hebei Food Inspection and Research Institute, Shijiazhuang 050091, China
* Correspondence: jiawenshen@163.com; Tel.: +86-13521217121

Abstract: To rapidly detect whether apples are infected by fungi, a portable electronic nose was used in this study to collect the gas information from apples, and the collected information was processed by smoothing filtering, data dimensionality reduction, and outlier removal. Following this, we utilized K-nearest neighbors (KNN), random forest (RF), support vector machine (SVM), a convolutional neural network (CNN), a back-propagation neural network (BPNN), a particle swarm optimization–back-propagation neural network (PSO-BPNN), a gray wolf optimization–backward propagation neural network (GWO-BPNN), and a sparrow search algorithm–backward propagation neural network (SSA-BPNN) model to discriminate apple samples, and adopted the 10-fold cross-validation method to evaluate the performance of each model. The results show that SSA can effectively optimize the performance of the BPNN, such that the recognition accuracy of the optimized SSA-BPNN model reaches 98.40%. This study provides an important reference value for the application of an electronic nose in the non-destructive and rapid detection of fungal infection in apples.

Keywords: electronic nose; fungal infection; sparrow search algorithm; apples

Citation: Zhao, C.; Ma, J.; Jia, W.; Wang, H.; Tian, H.; Wang, J.; Zhou, W. An Apple Fungal Infection Detection Model Based on BPNN Optimized by Sparrow Search Algorithm. *Biosensors* **2022**, *12*, 692. https://doi.org/10.3390/bios12090692

Received: 20 June 2022
Accepted: 17 August 2022
Published: 28 August 2022

1. Introduction

Apples are one of the most popular fruits, rich in vitamins and various minerals, and widely grown all over the world. Preliminary studies have shown that eating apples regularly can reduce the risk of colon cancer, prostate cancer, and lung cancer, while apple peel also contains a plethora of indeterminate phytochemicals that have antioxidant properties [1]. In 2021, the global production of apples was about 76.1 million tons, and normal apples can usually be stored for about a year after picking. However, apples are susceptible to some fungal infections during storage and transportation, including *Aspergillus niger* [2], *Penicillium expansum* [3] and *Penicillium crustosum* [4]. These fungi can cause widespread corruption of apples if not detected and handled in time, which will bring about huge direct economic losses for farmers. To reduce the economic loss caused by fungal infection, it is necessary to design a fast, convenient, and safe method to detect whether apples are infected by fungi, so as to improve the overall status of the apple industry.

When the volatile gas of apples encounters gas sensors, specific reactions occur, and the characteristic response spectra of gas information are given. Gas chromatography–mass spectrometry (GC-MS) is a widely used analytical method, which has the advantages of high selectivity, small sample requirement, and high resolution. GC-MS has developed rapidly in the applications of food safety, industrial detection, environmental protection, and other fields, while showing advantages, such as high resolution and high sensitivity.

Berrada et al. measured patulin in apple juice using GC-MS and investigated the effect of patulin on the stability of apples during storage [5]. Thin-layer chromatography (TLC) is a chemical analysis method that is widely used in qualitative analysis in the fields of food, medicine, and environment. TLC has a high detection efficiency and can complete an analysis in 10–20 min. High-performance liquid chromatography (HPLC) uses liquid as the mobile phase and uses a high-pressure system to effectively separate analytes. HPLC has the advantages of high sensitivity and a fast analysis speed and plays an important role in the field of food safety. It is also a globally recognized authoritative method for the qualitative detection of fungi [6]. However, although the above traditional methods can determine whether apples are infected by fungi, these processes need to be carried out in the laboratory, the detection process is complicated, operation by professionals is required, the cost of experiment is expensive, and real-time detection is not possible. Therefore, there is an urgent need for a fast and simple technique to determine whether the experimental sample is infected by fungi via the analysis of the volatile gas from the sample. The portable electronic nose is a fast, non-destructive, efficient, and chemical-free technology for analyzing volatile gas on-site [7]. In recent years, it has been used prominently, and is receiving increasing attention in the field of food safety and evaluation.

The convenient and fast operation of the electronic nose means it plays an important role in the quality inspection of fruits. Guo et al. collected the characteristic information of apples infected by different fungi through an electronic nose and used the BPNN pattern recognition model to classify and identify the information of apples; the results achieved were good [8]. Nouri et al. used an electronic nose combined with the BPNN pattern recognition model to detect the fungal infection of pomegranates, and the recognition rate of pomegranate samples infected with mycoplasma was as high as 100% [9]. This study shows that the electronic nose is a reliable and high-precision instrument for detecting the quality of pomegranate. Voss et al. used an electronic nose to capture the volatile gas of peaches to predict growth and maturity [10]. The results prove that the method of using an electronic nose combined with random forest (RF) can effectively predict the maturity date of peaches in orchards, thereby reducing farmers' economic losses caused by neglect and late harvest. Yang et al. analyzed the volatile gas in yellow peaches through an electronic nose, accomplished non-destructive prediction of the compression damage degree of the fruit, discriminated the damaged fruit, and predicted the compression time, while the accuracy of identifying the damaged fruit was as high as 93.33% [11]. Guo et al. used an electronic nose combined with PCA-DA to predict the corruption area and corruption degree of apples and achieved a good result. The prediction accuracy of the corruption degree was 97.2%; this study proves that the electronic nose has a certain application value in the classification of corrupt apples and the quantitative detection of corrupt areas [12]. To date, several studies have reported the application of the electronic nose technology in fruit quality detection. Furthermore, regarding the optimization of the recognition model, Wu et al. used a sparrow search algorithm (SSA) to propose an evaluation model for predicting the economic losses suffered by subway stations after rainstorms and floods, which effectively solves the problems of the low efficiency and low prediction accuracy of traditional evaluation models [13]. This study proves that the support vector machine (SVM) and BPNN evaluation models optimized by SSA have higher accuracy and stability than other optimization algorithms and can effectively predict the economic losses of subway stations caused by rainstorms and floods. Jiang et al. proposed a method for detecting aflatoxin B1 content in wheat based on colorimetric sensor array technology and used the firefly algorithm to optimize sensor features with SSA to optimize the BPNN recognition model [14]. The result proves that the prediction accuracy and stability of the optimized BPNN recognition model have been improved, and the complexity of the model has been reduced. Some studies have shown that the electronic nose can detect fruit quality according to volatile gas components under different conditions, which has great potential applicability for fruit damage and spoilage detection. Electronic nose technology can strengthen the early rot inspection of fruits and reduce the economic losses of growers.

However, few studies have used electronic noses combined with classification models to evaluate the effects of different fungi on apple spoilage. In addition, an electronic nose combined with the BPNN recognition model can often achieve better results; however, in the BPNN recognition model, the hidden layers are mainly responsible for modeling the complex functions of the network, and the number of nodes in the hidden layer has a great impact on the performance of the model, which may directly lead to the model falling into overfitting or underfitting. In the previous model building process, rich experience and continuous debugging were always required to find the appropriate number of nodes in the hidden layer. SSA is a new type of intelligent optimization algorithm that has been proposed in recent years; using SSA to optimize the number of nodes in the hidden layers of BPNN can help to quickly determine the appropriate number of nodes in each hidden layer, which precludes spending a lot of time and effort to debug the model manually while the performance of the optimized model is improved. We sought to detect conveniently, rapidly, and effectively fungal infection in apples, simplify the training process of the recognition model, and improve the performance of the recognition model. Therefore, this paper includes the following: (a) the use of the electronic nose to collect the volatile information of fresh apples, apples inoculated with *Aspergillus niger*, apples inoculated with *Penicillium expansum*, and apples inoculated with *Penicillium crustosum*; (b) the preprocessing of the collected data by filtering and removing outliers; (c) the dimensionality reduction of the preprocessed data; (d) finally, using the SSA-optimized BPNN model to classify apples infected with different fungi, comparing it with traditional pattern recognition methods.

2. Materials and Methods

2.1. Materials

The "Fuji" apples selected in this experiment came from apple plantations in Gansu Province, China. In total, 160 ripe apples were selected and randomly divided into 4 groups, 40 apples in each group, namely, Group A, Group B, Group C, and Group D. The fungi inoculated into the middle apples were *Aspergillus niger*, *Penicillium expansum*, and *Penicillium crustosum*. The apple samples were pretreated with 75% alcohol on a sterile bench and dried at room temperature. Then, four holes were punched in four directions in each apple in the three groups containing the inoculator (A, B, and C). Sample apples were inoculated with 7-day-old molds through drilled loops, and the holes were covered with sterile film. The mold-inoculated apples were then placed in a 1000 mL beaker, sealed with plastic wrap, and then placed in a 25 °C constant-temperature incubator for 5 days. Before the test, the apple samples were taken out of the incubator and left to rest for 30 min. To eliminate the influence of residual gas on the experimental results, the electronic nose was cleaned with inert gas before using. Electronic nose parameters were as follows: cleaning time 500 s, collection time 350 s, sampling interval 1 s, injection flow 150 mL/min.

The sensor array of the portable electronic nose in this experiment was composed of electrochemical sensors. The portable electronic nose was primarily composed of three parts: the control unit, the sensor room, and the data acquisition and transmission unit. The portable electronic nose used in the experiment is shown in Figure 1a, and the schematic diagram of the portable electronic nose is shown in Figure 1b. The sampling valve controls the gas entry into the sealed bottle, the injection valve controls the entry of gas into the air chamber and the flow of the gas, and the injection valve and the vacuum valve work together to prevent outside gas from entering the air chamber. The response curve of the No. 1 sensor C_2H_4-20 during the sampling process is shown in Figure 2. In addition, Fameview (V7.6.12.4) configuration software (Beijing Jiekong, Beijing, China) was used to collect and save the electronic nose data, while the Modbus protocol was used for data transmission and communication with the hardware, so as to permit human–computer interaction. In this experiment, two identical sensors (numbered 1 and 5) were selected as indicators to identify whether the collected data were abnormal. If the difference between the two 150–300 s sensors was greater than 1.2 mg/L, the specimen was considered anomalous and removed. Table 1 lists the sensor names and performance specifications, and Table 2 lists the specifications of the metal oxide sensors commonly used for PEN3 electronic noses. From the comparison between Tables 1 and 2, it can be seen that the recognition accuracy of the electrochemical sensor is higher than that of the metal oxide sensor. In addition, the electronic nose system proposed in this paper has a lower cost and a higher detection accuracy than the electronic nose of the same price. It is also very convenient to carry; the volume of the electronic nose system is about 0.04 m^3 and its weight is about 15 kg, which makes it convenient for inspection personnel as they carry out inspection and analysis on site. The operating platforms used in this experiment were PyCharm 2021.3 (JetBrains, Prague, Czech Republic), Tensorflow2 (Google, Menlo Park, CA, USA), and Matlab2018b (MathWorks, Portola Valley, CA, USA).

Table 1. Types of gas-sensitive sensors in the portable electronic nose sensor array, and their sensitive gas and detection accuracies.

Sensor Number	Sensor Model	Sensitive Gas	Detection Precision (ppm)
1	7NE/C_2H_4-20	C_2H_4	0.4
2	7NE/H_2S-50	H_2S	1
3	7NE/H_2S-1000	H_2S	20
4	7NE/C_2H_4-200	C_2H_4	4
5	7NE/C_2H_4-20	C_2H_4	0.4
6	7NE/ETO-20	C_2H_4O	0.4
7	7NE/PID-300	VOC	6
8	7NE/CH_2O-2000	CH_2O	40

Table 2. Names of sensors in the PEN3 electronic nose sensor array, their sensitive gas and detection accuracies.

Sensor Name	Sensitive Features	Representative Gas and Detection Precision (ppm)
W1C	Sensitive to aromatic compounds	Methylbenzene, 10
W3C	Aromatic compounds, particularly sensitive to ammonia	Benzene, 10
W5C	Aromatic compounds such as alkanes and compounds with relatively small polarity	Propane, 1
W1S	Particularly sensitive to methane contained in specimens	Methane, 100
W2S	Particularly sensitive to ethanol contained in specimens	Carbon monoxide, 100
W3S	Sensitive to high-concentration alkanes, especially methane, in specimens	Methane, 100
W5S	Sensitive to nitrogen oxides, extremely sensitive to negatively charged nitrogen oxides	Nitrogen dioxide, 1
W6S	Only detects hydrogen	Hydrogen, 100
W1W	Mainly sensitive to sulfides, also sensitive to organic sulfides	Hydrogen sulfide, 1
W2W	Mainly sensitive to aromatic compounds and organic sulfides	Hydrogen sulfide, 1

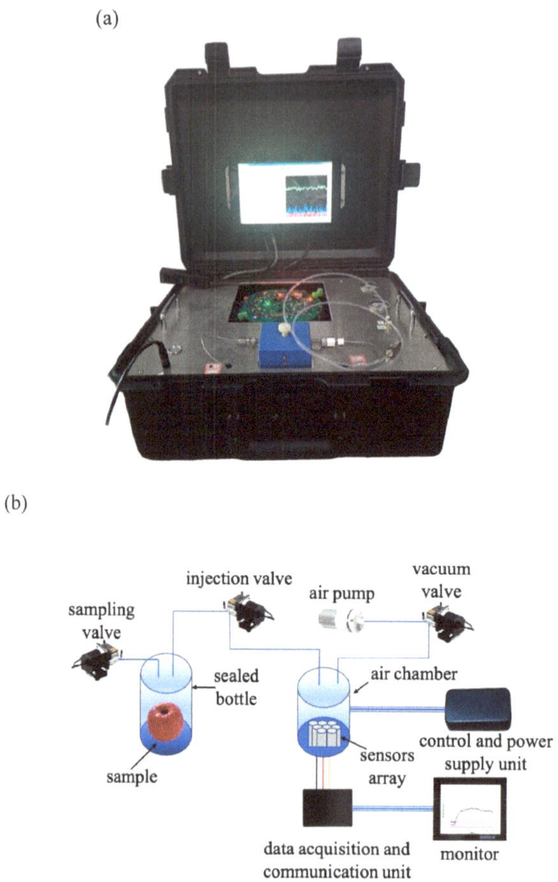

Figure 1. (**a**) The portable electronic nose; (**b**) schematic diagram of the portable electronic nose.

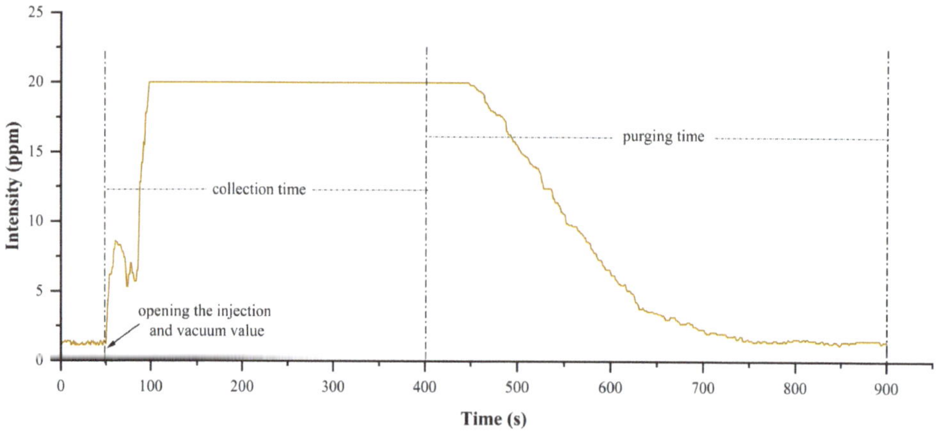

Figure 2. The response curve of the No. 1 sensor C_2H_4-20 during the entire sample-collection period. When opening the injection and vacuum valve at 50 s, the collection time lasts 350 s, and the purging of the electronic nose lasts 500 s.

2.2. Methods

2.2.1. Data Preprocessing

The portable electronic nose selected in this experiment is equipped with 8 sensors, and each sensor can detect different gas components. Since the sensor array of the portable electronic nose in this experiment used an electrochemical sensor, it has the characteristics of high sensitivity, but low response compared with metal oxide sensors, while the sensor has a certain cross-sensitivity (it can detect multiple gases). Based on this characteristic, we took the integral value, variance value, average differential value, maximum gradient value, relatively stable average value and energy value of the response curve of each sensor over 30–300 s as the characteristic information of the electronic nose, so the characteristic parameter of each sample is 48.

Smooth Filter

In this experiment, 3-point linear smoothing, 5-point linear smoothing, 7-point linear smoothing, 9-point linear smoothing, and 11-point linear smoothing algorithms were selected to remove noise from the data, and their results were compared. The response curves of each sensor after smoothing are shown in Figure 3. As can be seen from Figure 3, the response curve after 7-point smoothing is relatively smooth, and can maintain the basic characteristic information of the data, so the 7-point linear smoothing method was selected for the preprocessing operation in this experiment. In addition, the response curves of the 7NE/H_2S-1000 and PID-300 sensors are always 0, because fresh apples and apples infected with fungi release less H_2S gas. From Table 1, it can be seen that H_2S-50 and H_2S-1000 are both effective sensors for detecting H_2S gas; the difference between them is that the detection precision of H_2S-50 is much higher than that of H_2S-1000. It can be seen from Figure 3a that the response curve of the H_2S-50 sensor is less than 1, which means that the sample releases less H_2S gas, and has not reached the minimum detection range of H_2S-1000. Therefore, the response curve of H_2S-1000 is always 0. Similarly, the concentration of VOC gas released by the sample in the sealed bottle did not reach the minimum detection range of the PID-300 sensor, so the response curve of the PID-300 sensor is always 0. As such, in this study, we only analyzed the data measured by 6 sensors, excluding 7NE/H_2S-1000 and VOC-300, so the characteristic parameter of each sample is 36.

Eliminate Outliers

Due to the performances of the sensors and the effects of the external environment, the collected sample data may contain abnormal values, and abnormal data will directly affect the accuracy and stability of the recognition model. In order to eliminate this effect, Mahalanobis distance was used to remove abnormal data from the original data. Mahalanobis distance is a method used for calculating the distance between points and distribution and was proposed by Indian statistician P. C. Mahalanobis. The inconsistency and correlation between the scales of each dimension can be used to effectively evaluate the similarity within the data. Sun et al. used Mahalanobis distance combined with Monte Carlo cross-validation to effectively remove outliers from the hyperspectral data of tobacco leaf water content [15].

In this experiment, Mahalanobis distance and the method of difference judgment for the No. 1 and No. 5 sensors mentioned above were used to eliminate 30 abnormal sample data: 8 apples inoculated with *Penicillium crustosum*, 6 apples inoculated with *Aspergillus niger*, 5 apples inoculated with *Penicillium expansum*, and 11 fresh apples. Then, the KNN, SVM, and BPNN models were trained with the original data and the data after removing outliers, respectively, and the 10-fold cross-validation method was used to evaluate the performance of each model; the results are shown in Table 3. It can be seen that the accuracy and stability of each pattern recognition model were improved after removing outliers from Table 3. The results show that removing outliers from the raw data can effectively improve the performance of the recognition model.

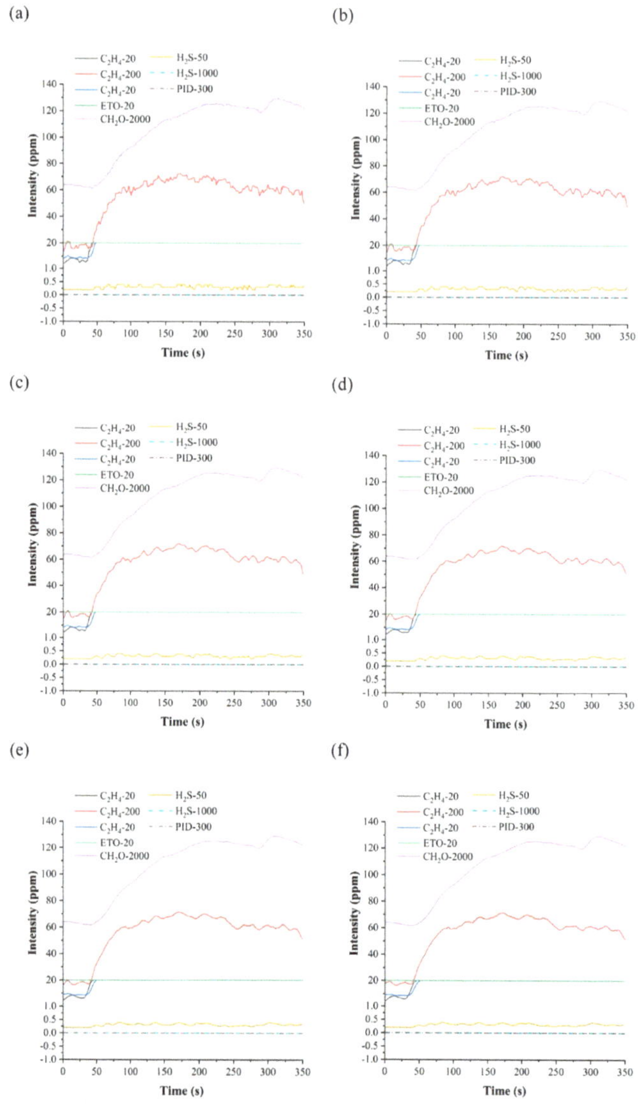

Figure 3. (**a**) Original response curve. (**b**) Response curve after 3-point smoothing filtering. (**c**) Response curve after 5-point smoothing filtering. (**d**) Response curve after 7-point smoothing filtering. (**e**) Response curve after 9-point smoothing filtering. (**f**) Response curve after 11-point smoothing filtering.

Table 3. Average accuracy and standard deviation of KNN, BPNN, and SVM pattern recognition models before and after removing outliers.

Model	Sample Set	Validation Set Accuracy	Standard Deviation of Accuracy
KNN	before removing	62.37%	0.326
	after removing	68.29%	0.163
BPNN	before removing	56.38%	0.452
	after removing	69.21%	0.263
SVM	before removing	57.32%	0.318
	after removing	64.65%	0.227

Data Dimensionality Reduction

Principal component analysis (PCA), factor analysis (FA), and linear discriminant analysis (LDA) are common data dimensionality reduction methods. The purpose of data dimensionality reduction is to reduce the dimension of the original data, remove useless information from the original data, and increase the recognition accuracy of the recognition model on the premise of retaining as much as possible of the feature information of the original data. PCA, FA, and LDA were performed on the sample data after removing outliers, and the results are shown in Figure 4. The 10-fold cross-validation method was used to evaluate the performances of the SVM, BPNN, and KNN recognition models with different dimensionality reduction methods. The results are shown in Table 4. The results show that the recognition accuracies of the models after PCA, FA, and LDA were improved, and the dimensionality reduction performance of LDA was the best among the three methods.

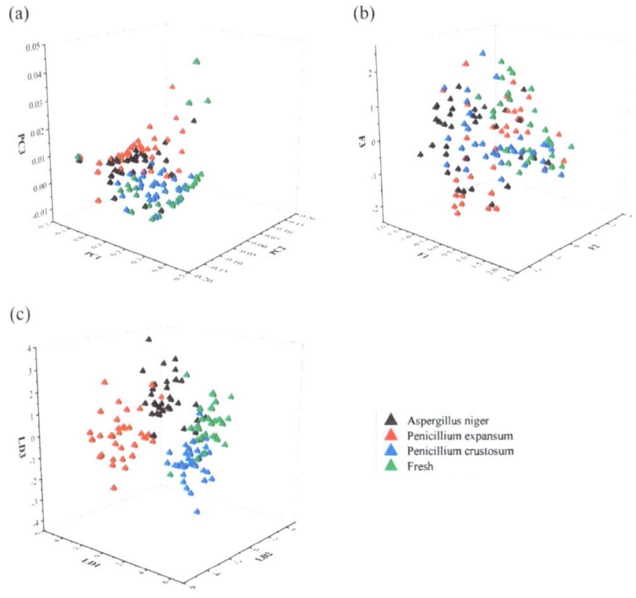

Figure 4. Comparison of different data dimensionality reduction methods: (**a**) PCA dimensionality reduction analysis; (**b**) FA dimensionality reduction analysis; (**c**) LDA dimensionality reduction analysis.

Table 4. Average accuracy of KNN, BPNN, and SVM pattern recognition models after different dimensionality reduction methods; the LDA dimensionality reduction method is the best among the three dimensionality reduction methods.

Dimensionality Reduction Method	Model	Average Accuracy	Standard Deviation of Accuracy
	SVM	64.65%	0.227
None	BPNN	69.21%	0.263
	KNN	68.29%	0.163
	SVM	66.29%	0.138
PCA	BPNN	75.38%	0.149
	KNN	73.85%	0.104
	SVM	65.50%	0.136
FA	BPNN	72.14%	0.147
	KNN	63.53%	0.119
	SVM	91.07%	0.046
LDA	BPNN	93.17%	0.072
	KNN	89.23%	0.058

2.2.2. Pattern Recognition Model

In recent years, with the rapid development of machine learning, the combination of electronic noses and pattern recognition models such as KNN, RF, SVM, CNN, and BPNN in machine learning has increased the prominence of the electronic nose in many fields. Virtanen et al. successfully identified five common pathogenic bacteria of acute sinusitis by combining the electronic nose and KNN while providing a pathological basis for the treatment of acute sinusitis [16]. Tian et al. made full use of the advantages of high RF stability, short time consumption, and high precision, and proposed an electronic nose and RF model based on the rapid detection of yogurt flavor acceptability. The research proved that the combination of an electronic nose and RF can be used to effectively evaluate the acceptability of yogurt flavor [17,18]. Jiang et al. used a combination of the electronic nose and SVM to classify five common odors [19]. Kang et al. uses CNN to process electronic nose data based on metal oxide sensor arrays to achieve the real-time detection of CO, NH_3, NO_2, CH_4, and C_3H_6O gases [20]. Gu et al. used a combination of the electronic nose and BPNN to detect early Aspergillus in rice, and the recognition accuracy reached 96.40% [21].

To improve the performance of the pattern recognition model, an optimization algorithm can be used. Currently, the most commonly used optimization algorithms are particle swarm optimization (PSO) [22], gray wolf optimization (GWO) [23], and sparrow search algorithm (SSA) [24]. These are inspired by the feeding behaviors of animals in nature. SSA is a recently proposed swarm intelligence optimization algorithm. The sparrow population in SSA is divided into two parts: producers and scroungers. Producers have a high fitness value and energy reserve, and their main task is to provide scroungers with directions and areas for foraging. The search range of producers is larger than that of scroungers. Scroungers follow producers to find food and obtain their own energy reserve, and thus increase their fitness value, and some scroungers continuously increase their own energy reserves through predation, thus turning themselves into producers. In addition, some sparrows in the sparrow population will act as forewarners. Forewarners will issue a warning signal when danger is coming, and at the same time spread into the safe area to obtain a better position. When the alarm value is greater than the set threshold, the producers will lead all scroungers out of the danger zone. A schematic diagram of SSA is shown in Figure 5. SSA has better global search and local development capabilities and can consider all the variable factors of the population, so that the population can quickly move into the optimal position. SSA also has the advantages of fewer iterations and higher prediction model accuracy.

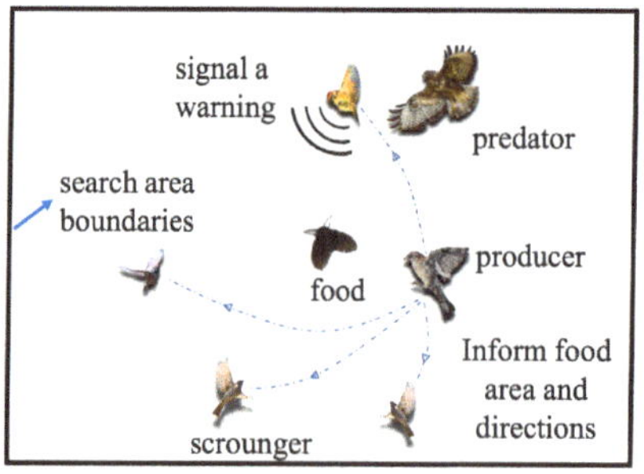

Figure 5. Schematic diagram of the sparrow search algorithm.

3. Results

In order to verify the abilities of the above eight recognition models to recognize fungus-infected apples, a multi-algorithm pattern recognition platform was developed using the PyQt5 tool in this experiment. The user interface enables human–computer interaction through the mouse and keyboard to adjust the parameters of different recognition models. As shown in Figure 6a, users can select different pattern recognition models on the main interface of the platform and enter the corresponding recognition model interface by clicking the button. Then, as shown in Figure 6b–i, the user can adjust the parameters of the current recognition model according to the label prompts, select training sample data to train the model, and assess the performance of the model. When the training of the recognition model is completed, the inspectors can begin to inspect the apple samples. First, one must put the apple sample into a sealed bottle to collect its electronic nose data, select the recognition model that has been trained in the multi-algorithm pattern recognition platform, and finally select the collected electronic nose data for this model to use to detect the apple sample. The result will soon appear in the recognition result display area below, as shown in Figure 6j. The whole process does not require the use of any chemical reagents, nor will it cause damage to the apple samples, and the testing process will not affect the edibility and sales of the apple samples.

Figure 6. *Cont.*

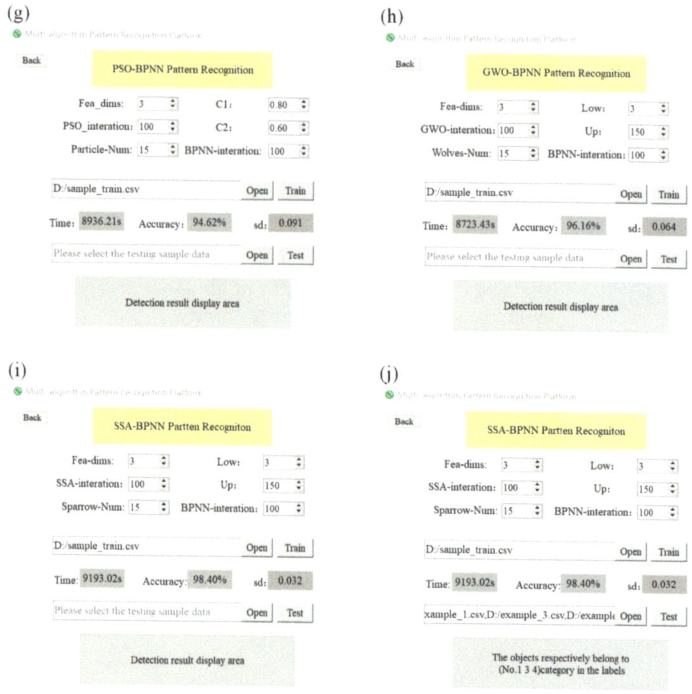

Figure 6. Multi-algorithm pattern recognition model platform. (**a**) Main interface of multi-algorithm pattern recognition platform. (**b**) KNN pattern recognition interface. (**c**) CNN pattern recognition interface. (**d**) SVM pattern recognition interface. (**e**) RF pattern recognition model interface. (**f**) BPNN pattern recognition model interface. (**g**) PSO-BPNN pattern recognition interface. (**h**) GWO-BPNN pattern recognition interface. (**i**) SSA-BPNN pattern recognition interface. (**j**) SSA-BPNN detection sample interface.

3.1. The Recognition Accuracy of the Model Optimized by SSA-BPNN Is Higher

The characteristic information of apples inoculated with three different fungi and fresh apples was collected by the electronic nose, and the collected characteristic information was preprocessed. Then, we used the above-mentioned multi-algorithm pattern recognition platform to train each recognition model while using the 10-fold cross-validation method to evaluate the performance of each model; the results are shown in Figure 6b–i, and the summarized results are shown in Table 5. It can be seen from Table 5 that CNN, RF, KNN, SVM, and BPNN were used for identifying the apples inoculated with *Aspergillus niger*, *Penicillium expansum*, and Penicillium officinale, as well as fresh apples. The average accuracy values of each recognition model are 57.80%, 86.92%, 89.23%, 91.07%, and 93.17%, and the standard deviation of the accuracy is less than 0.08. In terms of training time, although the training times of the RF, KNN, and SVM recognition models are shorter, they are not as good as the BPNN model in terms of recognition accuracy. Following this, three optimization algorithms, PSO, GWO, and SSA, were used to optimize the BPNN model. The number of iterations was set to 100 and the number of populations was 15; the accuracy of the BPNN model was used as the fitness function, and the 10-fold cross-validation method was also used to evaluate the performance of the model after optimization. Then, the average recognition accuracy values of the PSO-BPNN, GWO-BPNN, and SSA-BPNN models are 94.62%, 96.16% and 98.40%, respectively, and the standard deviations of the accuracy are 0.091, 0.064 and 0.032, respectively. The training times are 8936.21 s, 8723.43 s and 9193.02 s, respectively. In addition, as regards the evaluation score of the true positive rate (TPR) and the F1 score of each recognition model, the scores of SSA-BPNN are higher than those of other recognition models. In summary, the experimental results show that the SSA-BPNN recognition model proposed in this paper achieves outstanding performance in detecting fungal infection in apples.

Table 5. TPR, F1 score, average accuracy, standard deviation of accuracy, and training time of different pattern recognition models by 10-fold cross-validation.

Preprocessing	Model	TPR	F1 Score	Average Accuracy	Standard Deviation of Accuracy	Training Time (s)
\	CNN	61.03%	0.564	57.80%	0.071	270.16
LDA	RF	84.36%	0.836	86.92%	0.077	0.17
LDA	KNN	88.12%	0.878	89.23%	0.058	0.09
LDA	SVM	90.69%	0.903	91.07%	0.046	0.22
LDA	BPNN	92.75%	0.929	93.17%	0.072	63.12
LDA	PSO-BPNN	93.82%	0.937	94.62%	0.091	8936.21
LDA	GWO-BPNN	95.83%	0.951	96.16%	0.064	8723.43
LDA	SSA-BPNN	97.31%	0.976	98.40%	0.032	9193.02

3.2. SSA-BPNN Has Faster Convergence

The variations in the fitness functions of SSA, GWO, and PSO with the number of iterations in the optimization process are shown in Figure 7. Although SSA-BPNN has no obvious beneficial effect on the optimization time required for 100 iterations, it can be seen from Figure 7 that PSO-BPNN, GWO-BPNN, and SSA-BPNN enable the model to reach the optimal state after 81, 70, and 36 iterations, respectively, which shows that the convergence speed of SSA is higher than those of PSO and GWO, and it has a better optimization capacity. Compared with the GWO-BPNN and PSO-BPNN models, the SSA-BPNN model proposed in this study has obvious advantages in terms of recognition accuracy, stability, and convergence speed.

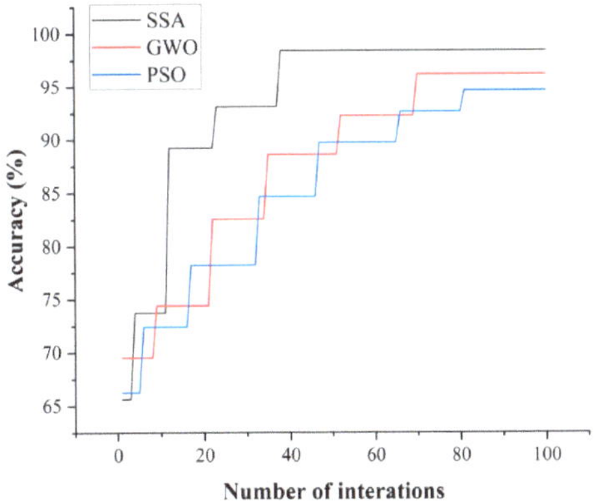

Figure 7. Variation trend of the fitness function curves of SSA, GWO, and PSO with the number of iterations. The SSA-BPNN model reached the optimal state first.

SSA is an intelligent optimization algorithm that has been proposed in recent years. Using SSA to optimize the BPNN recognition model can help the BPNN model to quickly find the optimal parameters thus avoiding the need to spend a lot of time and energy to debug the recognition model manually. With regard to the huge economic losses caused by fungal infection in the process of storage and transportation, the portable electronic nose combined with the SSA-BPNN method proposed in this study can effectively detect and identify common fungi and take preventive measures in time. The measures can effectively reduce the economic losses caused by fungal infection in apples.

4. Discussion

Apples are loved because of their delicious taste and rich nutrition. The annual global demand and supply of apples are huge, but some apples will inevitably be infected by fungi during the storage process, which will introduce huge economic losses to apple merchants. Electronic noses can be used to detect volatile substances in contact with the sensor array, which facilitates the quality detection of fruits. With the continuous development and progress of machine learning, the combination of electronic noses and machine learning provides a fast, non-destructive and easy-to-operate method for fruit quality detection and has gradually attracted people's attention. Compared with machine learning algorithms such as KNN, RF, CNN, and SVM, BPNN has the best performance. However, BPNN has many parameters of note; in particular, a small change in the number of nodes in the hidden layers can easily affect the overall performance of the BPNN network model. The selection of optimal parameters for BPNN has always been the most important task in the process of model building. On this basis, this paper proposed to use SSA to optimize the BPNN network model by finding the optimal parameters of the hidden layers of BPNN through continuous iteration, such that the model can quickly reach the optimal state. The TPR, F1 score, and accuracy of the model's recognition results after optimization achieved 97.31%, 0.976%, and 98.40%, respectively—4.56%, 0.047%, and 5.23% higher than those before optimization. The reason for this is the lower number of iterations compared to PSO and GWO. However, when the electronic nose is used to detect the quality of apples, its performance will be affected by external factors. Interference outside the normal range will give rise to abnormalities in the characteristic information of the samples collected by the electronic nose, which will eventually lead to unreliable test results. In order to improve the reliability of the detection results, in our follow-up research, we will combine the electronic nose with other detection equipment and collect characteristic information from the samples at the same time, using the method of data fusion to detect and identify the samples.

Author Contributions: Conceptualization, W.J., J.W. and J.M.; methodology, W.J. and C.Z.; software, C.Z.; validation, W.J., J.M., J.W. and C.Z.; formal analysis, W.J., J.M. and J.W.; investigation, H.T., W.Z. and H.W.; resources, W.J.; data curation, H.T., W.Z. and H.W.; writing—original draft preparation, C.Z.; writing—review and editing, W.J.; visualization, C.Z.; supervision, W.J. and J.M.; project administration, W.J.; funding acquisition, W.J. All authors have read and agreed to the published version of the manuscript.

Funding: This work was funded by the Hebei Province Key Research and Development Program under Grant No. 21375501D, the National Key Research and Development Program of China under Grant No. 2019YFC1605603, the Financial Supplementary Special Project of Beijing Academy of Agriculture and Forestry Sciences under Grant No. CZZJ202102, and the National Natural Science Foundation of China under Grant No. 31801634.

Institutional Review Board Statement: Not applicable.

Informed Consent Statement: Not applicable.

Data Availability Statement: The raw and processed data presented in this study are publicly available on Figshare at https://doi.org/10.6084/m9.figshare.19759120.v1 (accessed on 13 May 2022) and can be cited. The code in the study is publicly available on GitHub at https://github.com/palenn/Identifying_fungal-infected_apples.git (accessed on 13 May 2022) and can be cited.

Conflicts of Interest: The authors declare no conflict of interest.

References

1. Kang, N.J.; Lee, K.W.; Lee, S.J.; Lee, C.Y.; Lee, H.J. Effects of phenolics in Empire apples on hydrogen peroxide-induced inhibition of gap-junctional intercellular communication. *Biofactors* **2004**, *21*, 361–365. [CrossRef] [PubMed]
2. Munir, N.; Rafique, M.; Altaf, I.; Sharif, N.; Naz, S. Antioxidant and antimicrobial activities of extracts from selected algal species. *Bangladesh J. Bot.* **2018**, *47*, 53–61.
3. Guo, Z.; Wang, M.; Barimah, A.O.; Chen, Q.; Li, H.; Shi, J.; El-Seedi, H.R.; Zou, X. Label-free surface enhanced Raman scattering spectroscopy for discrimination and detection of dominant apple spoilage fungus. *Int. J. Food Microbiol.* **2020**, *338*, 108990. [CrossRef] [PubMed]
4. Žebeljan, A.; Vico, I.; Duduk, N.; Žiberna, B.; Krajnc, A.U. Profiling changes in primary metabolites and antioxidants during apple fruit decay caused by *Penicillium crustosum*. *Physiol. Mol. Plant Pathol.* **2020**, *113*, 101586. [CrossRef]
5. Berrada, H.; Buys, E.; Mañes, J.; Font, G. Determination of patulin in apple juice by GC–MS/MS: Stability study during storage. *Toxicol. Lett.* **2012**, *211*, S97. [CrossRef]
6. Zeng, W.; Tao, H.; Li, Y.; Wang, J.; Xia, C.; Li, S.; Wang, M.; Wang, Q.; Miao, H. The flavor of Chinese kale sprouts is affected by genotypic variation of glucosinolates and their breakdown products. *Food Chem.* **2021**, *359*, 129824. [CrossRef]
7. Liu, Y.; Furuno, S.; Akagawa, S.; Yatabe, R.; Onodera, T.; Fujiwara, N.; Takeda, H.; Uchida, S.; Toko, K. Odor Recognition of Thermal Decomposition Products of Electric Cables Using Odor Sensing Arrays. *Chemosensors* **2021**, *9*, 261. [CrossRef]
8. Guo, Z.; Guo, C.; Sun, L.; Zuo, M.; Chen, Q.; El-Seedi, H.R.; Zou, X. Identification of the apple spoilage causative fungi and prediction of the spoilage degree using electronic nose. *J. Food Process Eng.* **2021**, *44*, e13816. [CrossRef]
9. Nouri, B.; Mohtasebi, S.S.; Rafiee, S. Quality detection of pomegranate fruit infected with fungal disease. *Int. J. Food Prop.* **2020**, *23*, 9–21. [CrossRef]
10. Voss, H.G.J.; Ayub, R.A.; Stevan, S.L. E-nose Prototype to Monitoring the Growth and Maturation of Peaches in the Orchard. *IEEE Sens. J.* **2020**, *20*, 11741–11750. [CrossRef]
11. Yang, X.; Chen, J.; Jia, L.; Yu, W.; Wang, D.; Wei, W.; Li, S.; Tian, S.; Wu, D. Rapid and Non-Destructive Detection of Compression Damage of Yellow Peach Using an Electronic Nose and Chemometrics. *Sensors* **2020**, *20*, 1866. [CrossRef] [PubMed]
12. Guo, Z.; Guo, C.; Chen, Q.; Ouyang, Q.; Shi, J.; El-Seedi, H.R.; Zou, X. Classification for *Penicillium expansum* Spoilage and Defect in Apples by Electronic Nose Combined with Chemometrics. *Sensors* **2020**, *20*, 2130. [CrossRef] [PubMed]
13. Wu, H.; Wang, J. A Method for Prediction of Waterlogging Economic Losses in a Subway Station Project. *Mathematics* **2021**, *9*, 1421. [CrossRef]
14. Jiang, H.; Wang, J.; Mao, W.; Chen, Q. Determination of aflatoxin B1 in wheat based on colourimetric sensor array technology: Optimization of sensor features and model parameters to improve the model generalization performance. *Microchem. J.* **2022**, *175*, 107173. [CrossRef]
15. Sun, J.; Zhou, X.; Wu, X.; Zhang, X.; Li, Q. Identification of moisture content in tobacco plant leaves using outlier sample eliminating algorithms and hyperspectral data. *Biochem. Biophys. Res. Commun.* **2016**, *471*, 226–232. [CrossRef]
16. Virtanen, J.; Hokkinen, L.; Karjalainen, M.; Kontunen, A.; Vuento, R.; Numminen, J.; Rautiainen, M.; Oksala, N.; Roine, A.; Kivekäs, I. In vitro detection of common rhinosinusitis bacteria by the eNose utilising differential mobility spectrometry. *Eur. Arch. Oto-Rhino-Laryngol.* **2018**, *275*, 2273–2279. [CrossRef]
17. Ge, G.; Shi, Z.; Zhu, Y.; Yang, X.; Hao, Y. Land use/cover classification in an arid desert-oasis mosaic landscape of China using remote sensed imagery: Performance assessment of four machine learning algorithms. *Glob. Ecol. Conserv.* **2020**, *22*, e00971. [CrossRef]
18. Tian, H.; Liu, H.; He, Y.; Chen, B.; Xiao, L.; Fei, Y.; Wang, G.; Yu, H.; Chen, C. Combined application of electronic nose analysis and back-propagation neural network and random forest models for assessing yogurt flavor acceptability. *J. Food Meas. Charact.* **2020**, *14*, 573–583. [CrossRef]
19. Jiang, W.; Gao, D. Five typical stenches detection using an Electronic Nose. *Sensors* **2020**, *20*, 2514. [CrossRef]
20. Kang, M.; Cho, I.; Park, J.; Jeong, J.; Lee, K.; Lee, B.; Henriquez, D.D.O.; Yoon, K.; Park, I. High Accuracy Real-Time Multi-Gas Identification by a Batch-Uniform Gas Sensor Array and Deep Learning Algorithm. *ACS Sens.* **2022**, *7*, 430–440. [CrossRef]
21. Gu, S.; Wang, J.; Wang, Y. Early discrimination and growth tracking of *Aspergillus* spp. contamination in rice kernels using electronic nose. *Food Chem.* **2019**, *292*, 325–335. [CrossRef] [PubMed]
22. Eberhart, R.; Kennedy, J. A new optimizer using particle swarm theory. In Proceedings of the Sixth International Symposium on Micro Machine and Human Science, MHS'95, Nagoya, Japan, 4–6 October 1995; IEEE: New York, NY, USA, 1995; pp. 39–43.
23. Mirjalili, S.; Mirjalili, S.M.; Lewis, A. Grey wolf optimizer. *Adv. Eng. Softw.* **2014**, *69*, 6–61. [CrossRef]
24. Xue, J.; Shen, B. A novel swarm intelligence optimization approach: Sparrow search algorithm. *Syst. Sci. Control. Eng.* **2020**, *8*, 22–34. [CrossRef]

biosensors

MDPI

Article

Variational Mode Decomposition Weighted Multiscale Support Vector Regression for Spectral Determination of Rapeseed Oil and Rhizoma Alpiniae Offcinarum Adulterants

Xihui Bian [1,2,3,*], Deyun Wu [1], Kui Zhang [1], Peng Liu [1], Huibing Shi [3], Xiaoyao Tan [1] and Zhigang Wang [1]

1 State Key Laboratory of Separation Membranes and Membrane Processes, School of Chemical Engineering and Technology, Tiangong University, Tianjin 300387, China; wudeyun1221@163.com (D.W.); zhangkui@dicp.ac.cn (K.Z.); liupeng012@tiangong.edu.cn (P.L.); tanxiaoyao@tiangong.edu.cn (X.T.); wangzhigang@tiangong.edu.cn (Z.W.)
2 State Key Laboratory of Plateau Ecology and Agriculture, Qinghai University, Xining 810016, China
3 Shandong Provincial Key Laboratory of Olefin Catalysis and Polymerization, Shandong Chambroad Holding Group Co., Ltd., Binzhou 256500, China; huibing.shi@chambroad.com
* Correspondence: bianxihui@mail.nankai.edu.cn

Abstract: The accurate prediction of the model is essential for food and herb analysis. In order to exploit the abundance of information embedded in the frequency and time domains, a weighted multiscale support vector regression (SVR) method based on variational mode decomposition (VMD), namely VMD-WMSVR, was proposed for the ultraviolet-visible (UV-Vis) spectral determination of rapeseed oil adulterants and near-infrared (NIR) spectral quantification of rhizoma alpiniae offcinarum adulterants. In this method, each spectrum is decomposed into K discrete mode components by VMD first. The mode matrix Uk is recombined from the decomposed components, and then, the SVR is used to build sub-models between each Uk and target value. The final prediction is obtained by integrating the predictions of the sub-models by weighted average. The performance of the proposed method was tested with two spectral datasets of adulterated vegetable oils and herbs. Compared with the results from partial least squares (PLS) and SVR, VMD-WMSVR shows potential in model accuracy.

Keywords: variational mode decomposition; support vector regression; adulteration; quality control; chemometrics

Citation: Bian, X.; Wu, D.; Zhang, K.; Liu, P.; Shi, H.; Tan, X.; Wang, Z. Variational Mode Decomposition Weighted Multiscale Support Vector Regression for Spectral Determination of Rapeseed Oil and Rhizoma Alpiniae Offcinarum Adulterants. *Biosensors* **2022**, *12*, 586. https://doi.org/10.3390/bios12080586

Received: 9 July 2022
Accepted: 27 July 2022
Published: 1 August 2022

1. Introduction

Quality control is a critical analytical topic, especially regarding foods and herbs that play an essential role in everyday life. Adulteration is one of the major challenges in the quality control of foods and herbs. Some unscrupulous vendors dilute them with cheap alternative food sources or fraudulently label and sell low-quality products as premium ones to make more profit [1]. These not only damage the quality and nutritional value of foods but also are detrimental to consumer health. With a series of food adulteration incidents per year that lead to severe health impacts and economic costs, the problem of food fraud has become more sinister and devastating in the globalized food supply chain [2–4]. An important index for evaluating the nutritional value and quality of foods and herbs is to determine the contents of the main components in them. However, it is difficult to determine adulterants in the final product as some foods or herbs are similar in appearance but vary greatly in price [5,6]. In order to protect consumer rights and ensure food safety, there is an urgent need to develop a simple and reliable method to meet the accurate quantitative analysis of food and herb adulteration.

Various techniques have been applied to detect adulteration in foods and herbs, such as chromatography, mass spectrometry and capillary electrophoresis. These methods allow for a relatively accurate determination of the samples [7,8]. However, most of them are time-consuming and expensive and require a high degree of technical expertise [9]. Spectroscopic techniques, especially ultraviolet-visible (UV-Vis) and near-infrared (NIR) spectroscopy, have been rapidly developed in scientific research and industrial production because of their non-contact, environmentally friendly and low-cost advantages [10–13]. Since the original spectra usually contain a large amount of signal overlap, background and noise information unrelated to the target, chemometric models need to be integrated to improve and expand the potential applications of the spectroscopic techniques.

Chemometrics, as an effective support means, has been developed extensively in analytical chemistry [14–17], especially in multivariate calibration methods for spectral data analysis, such as partial least squares (PLS) and support vector regression (SVR) [18,19]. PLS is a commonly used modeling method because of its practicality and versatility, but it may produce undesirable prediction results when dealing with strongly nonlinear issues [18]. SVR has the capability to solve both linear and nonlinear multivariate regression problems with a simple process [20]. These modeling approaches predict unknown samples by constructing one model, but the prediction performance of only a single model that is built between spectra and targets tends to be poor when the training set is small or the samples are outliers [21].

Ensemble modeling has gained increasing attention in the multivariate calibration for quantitative analysis [22,23]. Compared with the prediction of a single model, ensemble modeling achieves a greater accuracy and more robust results by combining the predictions of multiple sub-models to produce the final prediction [21]. One of its key points is the generation of training sub-sets that can be produced from samples, variables or both directions, such as bagging, cluster and boosting [24–26]. However, most spectra are essentially localized and have varying localization in time and frequency. These traditional ensemble strategies are all generated sub-models from the original data that do not use both time and frequency information of the signal simultaneously [27]. Due to the complexity of the spectra, if the original signal is decomposed by mathematical transformation before ensemble calibration, better results may be obtained. There is different information hidden in the data that can be revealed by converting signals from the original data space to other spaces through a certain mathematical transformation.

Three decomposition strategies are widely applied for the signal process, that is, the Fourier transform (FT) [28], wavelet transform (WT) [29] and empirical mode decomposition (EMD) [30]. FT portrays well the frequency domain information of signals, but it does not provide time domain information and can only deal with stationary and linear signals [31]. WT has displayed its modeling effectiveness owing to its capacity for time-frequency resolutions. Nevertheless, WT is not a self-adaptive decomposition that needs to choose wavelet filters and scales for a given application to obtain an optimal result [32]. EMD is a useful technique for processing non-stationary and nonlinear signals and decomposes the signal into a finite number of intrinsic mode functions (IMFs) [30]. Although this self-adaptive decomposition method is a potent tool for the multiscale analysis of data without the trouble of selecting the filters or scales, the existence of mode mixing and end effect in the EMD process will lead to the distortion of IMF components [33]. Therefore, it is necessary to develop a new mathematical transformation for the signal process, which can make up for the deficiency in the above methods.

Variational mode decomposition (VMD) is a new adaptive signal decomposition strategy that is particularly suitable for nonlinear and non-stationary signals [34]. It not only has a good separation effect on the noise in signals but also effectively suppresses the mode mixing and end effect [35]. Using VMD, a series of mode components can be decomposed from the complex spectra according to the inherent characteristics of the signals. Previously, various studies reported that VMD has been used successfully in multiple fields due to its efficiency superiority, such as the forecast of stock prices [36], wind speed forecasting [37] and fault diagnosis [38]. However, there are very few reports in the literature that use the VMD algorithm for ensemble modeling in the spectral determination of food. Since VMD can fully utilize the information embedded over the frequency and time domains of spectral signals, it was introduced in the generation of sub-models.

Herein, a weighted multiscale SVR modeling method based on VMD for improving the prediction accuracy of food and herb adulterants is proposed and referred to as VMD-WMSVR. Firstly, each spectral signal is decomposed by VMD and then K mode components with different central frequencies are obtained. After recombining these mode components, SVR is used to establish sub-models for each mode. Finally, the predictions of each sub-model are weighted and averaged to obtain the ultimate prediction result. The spectral datasets of adulterated vegetable oils and herbs were investigated using this method. The performance of the method was evaluated based on the root mean square error of prediction (RMSEP) and correlation coefficient (R) and compared with results derived from single PLS and SVR models.

2. Materials and Methods

2.1. Sample Preparation

For adulterated vegetable oils, the sample consists of six different vegetable oils bought in different markets in the municipality of Tianjin. These are sesame oil, soybean oil, corn oil, peanut oil, rapeseed oil and sunflower oil. The six pure oils were blended in different mass proportions in order to form 51 adulterated vegetable oil samples. Each oil content is within the range of 0–100% (g/g) with an interval of ca. 2%. Before measurement, these samples were well shaken and sonicated in an ultrasonic instrument (SK6200HP, Kudos Ultrasonic Instrument Company, Shanghai, China) for 30 min to further mix and eliminate air bubbles. In this study, rapeseed oil was taken as the analysis target.

For adulterated herbs, the sample includes pure herbs. These are *Panax notoginseng* (PN), rhizoma alpiniae offcinarum (RAO), rhizoma curcumae (RC) and *Curcuma longa* (CL) purchased from various pharmacies in Tianjin. Since the herbs have a certain amount of moisture, they were dried at 60 °C to a constant weight. These herbs were ground into powder, passed through a 120-mesh stainless steel sieve and stored in sealed plastic bags measuring 60 mm × 100 mm. The four processed herbs were mixed at different mass percentages and ensured that the total mass fraction of the four herbs in each sample was 100%. There were 75 samples in the adulterated herbs dataset and studied with the content of RAO.

2.2. Spectral Collection

Two small spectral datasets were experimentally investigated. A UV-Vis spectrophotometer (Evolution 300, Thermo Fisher, Waltham, MA, USA) was used for the adulterated vegetable oils in order to obtain the spectra of 51 samples in the wavelength range from 200 to 800 nm with an interval of 1 nm. The average spectrum of three parallel measurements was used for each sample. There is a negative absorbance for the 200–380 nm wavelengths, which seems to have no useful information. When the absorbance is above four, there is an obvious noise phenomenon and the absorption peaks are mainly present at 380–800 nm. Thus, Figure 1a mainly shows the absorption peaks at wavelengths of 380–800 nm. The adulterated herbs were measured from 12,000 to 4000 cm^{-1} at 2 cm^{-1} intervals on a Vertex 70 NIR spectrometer (Bruker Optics Inc., Ettlingen, Germany). Figure 1b shows the NIR spectra of the samples.

Each spectrum of the adulterated samples in the same dataset is similar. Therefore, it is necessary to combine multivariate calibration with spectroscopy to achieve an accurate quantitative analysis. Before calculation, the two datasets were divided into the training and prediction set by the Kennard–Stone (KS) algorithm. KS is the most widely used grouping method in chemometrics, which usually yields good grouping results. For the vegetable oil dataset, 34 and 17 samples are used as the training and prediction sets, respectively. For the herb dataset, 50 and 25 samples are used as the training and prediction sets, respectively.

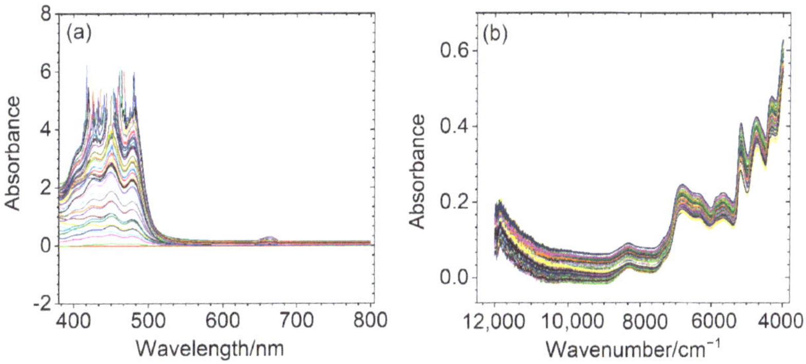

Figure 1. Measured spectra for adulterated vegetable oil (**a**) and herb (**b**) datasets.

2.3. Variational Mode Decomposition (VMD)

VMD is a powerful technique for signal analysis, which depends on the frequency information of the signal. The basic idea of the VMD algorithm is to construct and solve variational problems. For the construction of the variational problem, the purpose of VMD is to decompose the spectral signal **X** into a number of K discrete mode components u_k around the center frequency ω_k. At the same time, the sum of each mode is equal to the input signal **X**. The constrained variational model consists of the following target function.

$$\begin{cases} \min\limits_{\{u_k,\ \omega_k\}} \left\{ \sum\limits_k \left\| \partial_t \left[\left(\delta(t) + \frac{j}{\pi t} \right) * u_k(t) \right] e^{-j\omega_k t} \right\|_2^2 \right\} \\ s.t. \sum\limits_k u_k = \mathbf{X} \end{cases} \tag{1}$$

where $\{u_k\} = \{u_1, \ldots, u_K\}$ is the mode ensemble obtained by decomposition, $\{\omega_k\} = \{\omega_1, \ldots, \omega_K\}$ represents the center frequency of each mode component, δ is the Dirac function, $||\cdot||_2$ is the L2 distance, $*$ is the convolution, j is the imaginary unit and **X** is a $[m \times n]$ matrix containing n spectral responses of m samples.

By introducing Lagrange multipliers and quadratic penalty terms, the above problem can be transformed into an unconstrained variational problem. An alternate direction method of multipliers (ADMM) is used to solve the saddle points of the multipliers' function. $\{u_k\}$, $\{\omega_k\}$ and the Lagrange multiplier are updated continuously in the frequency domain until the optimal solution of the variational problem is obtained. Finally, the results are derived by a FT. Please refer to Ref [34]. for the detailed algorithm.

2.4. Support Vector Regression (SVR)

SVR is a machine learning algorithm based on the principle of structural risk minimization and function approximation. It is specifically used to obtain predictive models via a number of identified support vectors and nonlinear kernel functions. The main process of SVR is to map the input data into a high-dimensional space by kernel functions. Then, the optimal hyper-plane is found in this feature space and a model is built to solve the linear regression problem. With strict statistical theory, SVR is able to be trained with few samples. Least square SVR (LSSVR) is one of the SVR algorithms. It can transform the quadratic programming problem into the problem of solving linear equations to reduce the complexity of computation. The Lagrangian function that is constructed to solve the linear system is as follows:

$$\begin{bmatrix} 0 & \mathbf{I}_n^{\mathrm{T}} \\ \mathbf{I}_n\mathbf{K} + \gamma^{-1}\mathbf{I} \end{bmatrix} \begin{bmatrix} b_0 \\ \mathbf{b} \end{bmatrix} = \begin{bmatrix} 0 \\ y \end{bmatrix} \tag{2}$$

where $\mathbf{I}_n$ is a $[n \times 1]$ vector, **K** is a $[n \times 1]$ kernel matrix, T is a transpose of a matrix or vector, γ is a weight vector, **b** is regression vector and b_0 is the model offset.

In this study, the Gaussian radial basis function (RBF) kernel function was used:

$$k_{i,j} = e^{\frac{-|\mathbf{x}_i - \mathbf{x}_j|^2}{2\sigma^2}} \quad (3)$$

where $\mathbf{x}_i$ and $\mathbf{x}_j$ denote the measured spectra of different samples and σ is the kernel width parameter. As we can see from Equations (2) and (3), the performance of the SVR model is mainly affected by two parameters, namely, γ and σ^2. More details are provided in Refs [22,39].

2.5. Variational Mode Decomposition Weighted Multiscale Support Vector Regression (VMD-WMSVR)

Motived by the advantages of VMD and SVR, a novel ensemble modeling method (VMD-WMSVR) is proposed for the spectral quantitative analysis of food and herb adulterants. This method includes the calibration and prediction stages. The schematic diagram of the proposed method is shown in Figure 2. Details of the process are described as follows.

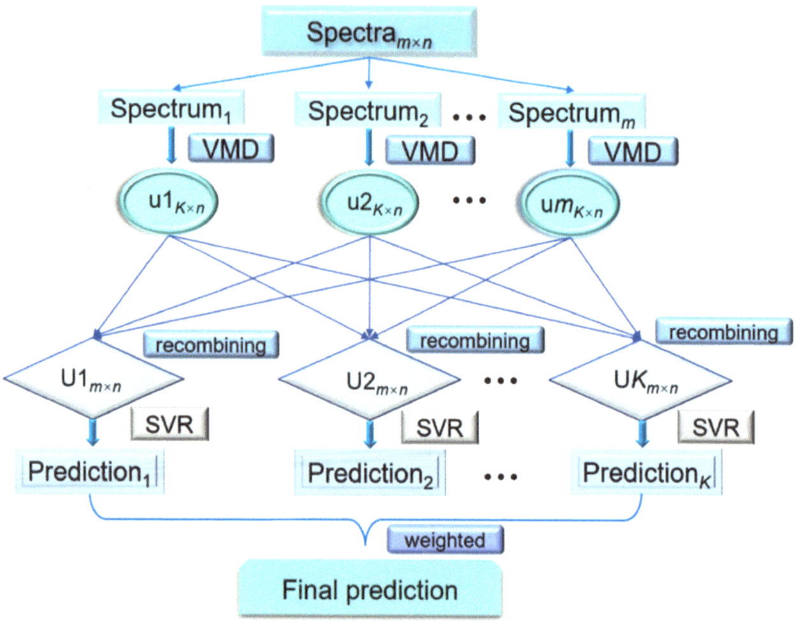

Figure 2. The schematic diagram of VMD-WMSVR.

(1) Each spectrum of the training set is decomposed by VMD into K discrete mode components uk (k = 1, 2, ..., K). Different K values of different samples have a large impact on the predictive stability of the proposed model. Moreover, too many mode components may destroy the linear relationship between the signal and the target value. The mode number K needs to be predetermined.

(2) Then, K mode components uk of the ith (i = 1, 2, ..., m) spectral signal are assigned sequentially to the ith row of each corresponding mode matrix Uk, i.e., the mode components uk are recombined to derive K modes Uk. In this way, each Uk contains the same number of samples and variables as the training set.

(3) SVR is used to build sub-models between each Uk and the target values. Overall, K multiscale regression sub-models are established for calibration.

(4) In the process of prediction, the spectral decomposition and recombination of the prediction set is the same as that of the training set. VMD-WMSVR is used for predicting

the samples in prediction set. Each sub-model gives a prediction and all the predictions are weighted and averaged to obtain the final prediction result. Sub-model weights are the inverse of the fourth power of the root mean square error of cross-validation (RMSECV).

3. Results and Discussion

3.1. The Mode Number K

In VMD-WMSVR model, the mode number *K* is a key parameter that needs to be set before the algorithm runs. Too few numbers of *K* may cause multiple components of the signal to be contained in one mode concurrently, resulting in insufficient decomposition. The model with too many numbers of *K* will create over decomposition problems and false modes [40]. In order to obtain the proper *K* value, the variation in the RMSEP with the mode number *K* for the two datasets is presented in Figure 3a,b, respectively.

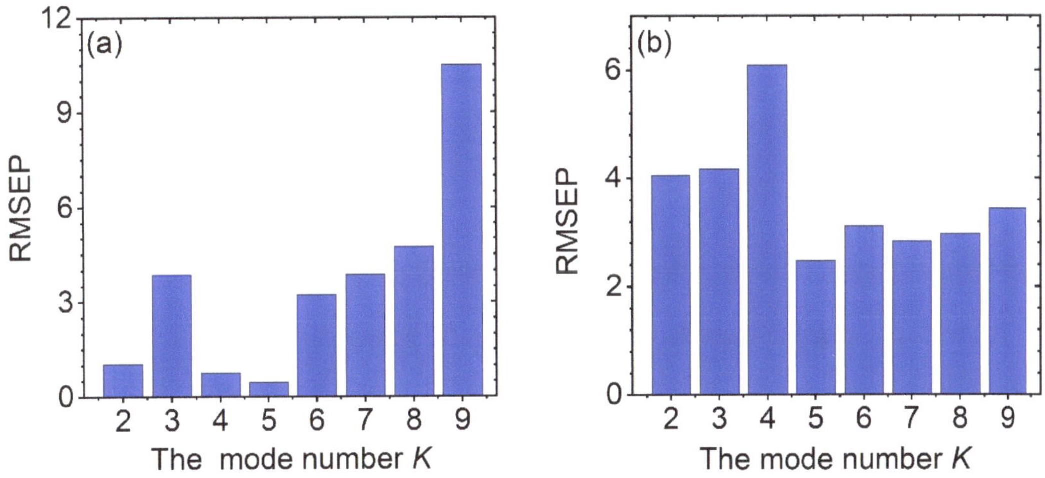

Figure 3. Variation in the RMSEP of VMD-WMSVR modeling with the mode number *K* for the adulterated vegetable oil (**a**) and herb (**b**) datasets.

VMD was applied multiple times with different *K* values for each spectrum and the result of the minimum RMSEP was considered the appropriate *K* value. Figure 3a shows that the RMSEP first presents a downward trend with the increase in *K*. When the *K* value reaches 5, the RMSEP obtains the minimum value and the prediction accuracy is the highest. After that, the RMSEP increases, especially when *K* is 9. This indicates that it may have mode mixing or pure noise modes that do not contribute much to the target value of interest. Figure 3b shows a similar trend to Figure 3a. As *K* is 5, the RMSEP reaches its minimum value. The smaller the RMSEP, the higher predictive accuracy of the model. Hence, the mode number *K* was set to 5 for both datasets.

3.2. The Spectral Decomposition of VMD

With the determination of *K*, each spectrum of the training set is decomposed by VMD and obtains *K* discrete mode components u*k* using this method. The UV-Vis spectra of 34 samples for the adulterated vegetable oils in the training set are decomposed. To illustrate the decomposition result, sample No. 2 is used. Figure 4a demonstrates that the original UV-Vis spectra are decomposed into five u components, which are graphically explained in the extracted order. This order represents the change in frequency from the lowest frequency to the highest. Different frequency blocks may contain different information and contribute varyingly to the model. The first three u components fluctuate slightly with a low number of peaks and the wavelength fluctuation in the range of 560–800 nm is gentle, which may contain some useful information. For u4, it almost fluctuates symmetrically near zero over the entire wavelength range by detailed observation. In addition, big peaks alternate with small ones, making it difficult to determine whether they are noise or not. The variation in frequency for u5 is higher and the pronounced peak number increment is observed compared with the former u components, which behave more like noise.

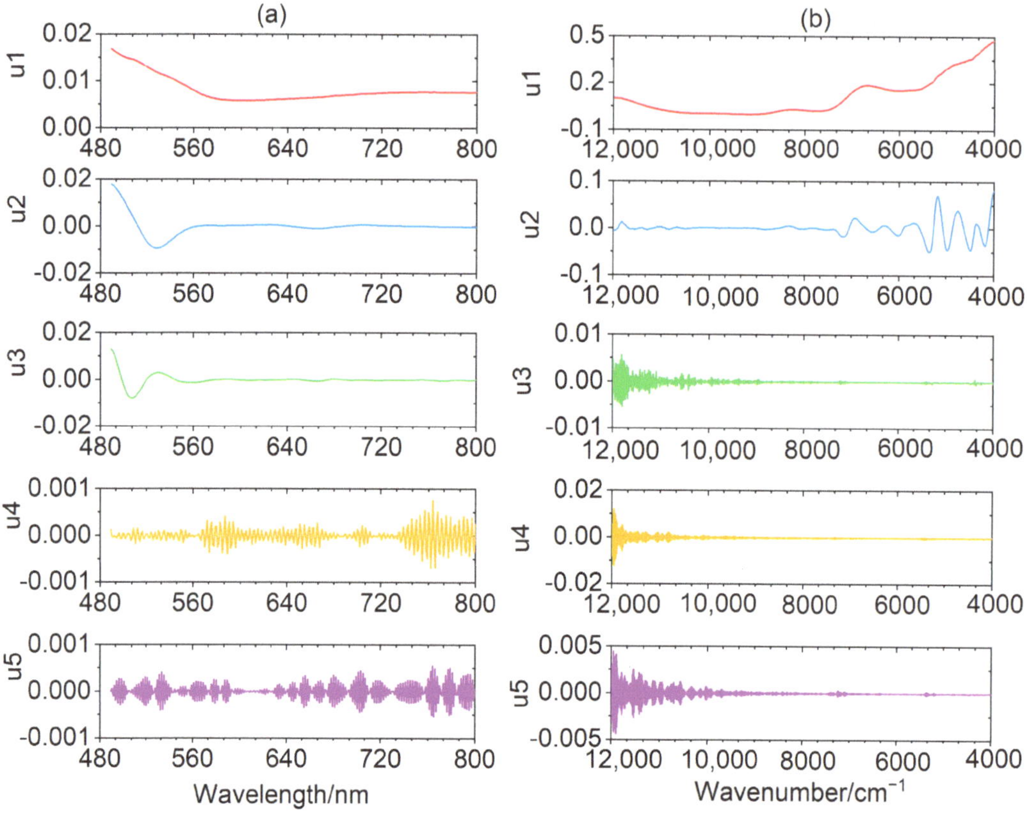

Figure 4. VMD diagram of the spectrum for sample No. 2 in the adulterated vegetable oil (**a**) and sample No. 17 in the adulterated herb (**b**) datasets.

For the adulterated herb dataset, each NIR spectrum of 50 samples in the training set is decomposed. Sample No. 17 is taken as an example. Figure 4b shows u1 oscillates slowly over the whole wavenumber with few peaks. Compared with u1, u2 changes more frequently and contains more peaks. Both of them have minor changes between 12,000 and 8000 cm^{-1} and have big fluctuations between 8000 and 4000 cm^{-1}, which may include a lot of helpful information. The last three u components have similar trends throughout the wavenumber range, fluctuating almost symmetrically around zero. The variation in frequency is prominent, changing rapidly from 12,000 to 10,000 cm^{-1}, which may have noise interference. In short, it can be seen from Figure 4 that the first u is a low-frequency component with a clear linear characteristic and a highly noticeable trend. As the order increases, the u component frequency becomes higher and higher, appearing with more irregularities and higher degrees of complexity. Since VMD is a mathematical decomposition, not all mode components have a well-defined chemical meaning for the spectral signal. The low-frequency and high-frequency mode components can be distinguished by observing their variation regularity combined with their variance values [41]. The low-frequency mode component changes gently with a large variance, while the high-frequency mode component oscillates almost symmetrically at zero with a small variance.

3.3. Comparison of the Predicted Results

In order to evaluate the predictive ability of the proposed method, PLS and SVR are used for comparison. The parameters of PLS and SVR were optimized at first. For PLS, the Monte Carlo cross validation (MCCV) combined with the F-test was used to determine latent variables (LVs). The optimal LV for the adulterated vegetable oil and herb datasets is 5 and 4, respectively. For SVR, there are two parameters (γ, σ^2) that need to be predetermined. The particle swarm optimization (PSO) algorithm was adopted and the RMSEP was used as the evaluation standard for the parameters optimization. The optimal γ and σ^2 for the adulterated vegetable oil and herb datasets are 222.74, 227.37 and 247.46, 106.22, respectively. The relationship between the prepared and the predicted values for the prediction set by PLS (a), SVR (b) and VMD-WMSVR (c) for the two adulterated datasets is shown in Figure 5andFigure 6, respectively. The RMSEP and R of the prediction set are used as indicators to validate the performance of the models.

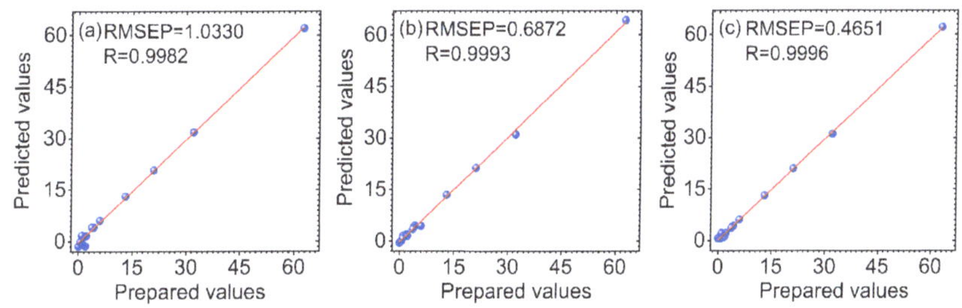

Figure 5. The relationship between the prepared and the predicted values for the prediction set by PLS (**a**), SVR (**b**) and VMD-WMSVR (**c**) for the adulterated vegetable oil dataset.

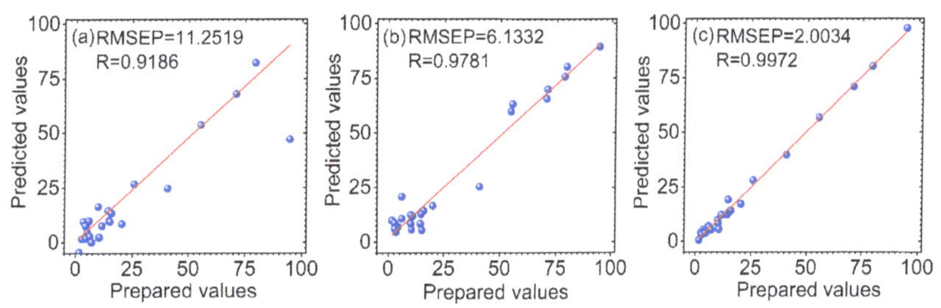

Figure 6. The relationship between the prepared and the predicted values for the prediction set by PLS (**a**), SVR (**b**) and VMD-WMSVR (**c**) for the adulterated herb dataset.

It can be found that the R values of the three methods are all above 0.9, indicating that these modeling methods combined with spectroscopy are effective for the quantitative analysis of rapeseed oil and RAO adulterants. However, the high benchmark of R leads to little room for improvement in all approaches, so their improvement is not significant from the perspective of R values alone. The variation in the RMSEP was used as the main criterion for comparison of the different methods. The RMSEP is a measure of the deviation between the predicted and prepared values. The smaller its value, the closer the predicted value is to the prepared value. For the vegetable oil dataset, it is observed from Figure 5 that SVR has a lower RMSEP and a higher R compared with PLS, demonstrating that SVR is superior to PLS. Among the three methods, VMD-WMSVR has the lowest RMSEP and the highest R. This indicates that the adaptive spectral decomposition can further improve the prediction ability of SVR and PLS. Thus, VMD-WMSVR has the best prediction, which is attributed to the original spectra of VMD. Figure 6 shows that compared with PLS, the RMSEP for the herb dataset under SVR is reduced by 45%. This also indicates that SVR is better than PLS in modeling results. There is a good linearity between the prepared values and the predicted values in Figure 6c. Compared with PLS and SVR, the RMSEP of VMD-WMSVR is reduced by 82% and 67%, respectively. Therefore, the prediction results of both datasets suggest the potential of VMD-WMSVR in improving the predictive accuracy.

4. Conclusions

In summary, this work presented a new chemometric methodology named VMD-WMSVR and was applied for two spectral datasets to achieve the quantification of vegetable oil and herb adulterants. On the one hand, VMD is designed to make full use of the information by decomposing the original spectra adaptively into multiple mode components with different frequencies. The modeling technique can improve the accuracy of predictions compared with a single PLS and SVR. On the other hand, it is a non-destructive and efficient method for the determination of rapeseed oil and RAO adulterants without the use of reagents and the generation of harmful residues, which protects the environment. However, the performance of predicting actual new samples was not discussed in this paper and should be further studied in the future.

Author Contributions: Conceptualization, X.B. and D.W.; methodology, X.B.; software, X.B., D.W. and K.Z.; validation, X.B. and Z.W.; investigation, X.B., K.Z. and D.W.; resources, X.B., H.S. and X.T.; data curation, X.B., D.W. and P.L.; writing—original draft preparation, X.B. and D.W.; writing—review and editing, X.B., Z.W. and D.W.; funding acquisition, X.B., H.S. and X.T. All authors have read and agreed to the published version of the manuscript.

Funding: This research was funded by the China Scholarship Council (No. 201808120028), Tianjin Science and Technology Program (No. 21ZYJDJC00100) and the Opening Foundation of State Key Laboratory of Plateau Ecology and Agriculture (No. 2021-KF-07).

Institutional Review Board Statement: Not applicable.

Informed Consent Statement: Not applicable.

Data Availability Statement: Not applicable.

Conflicts of Interest: The authors declare no conflict of interest.

References

1. Nehal, N.; Choudhary, B.; Nagpure, A.; Gupta, R.K. DNA barcoding: A modern age tool for detection of adulteration in food. *Crit. Rev. Biotechnol.* **2021**, *41*, 767–791. [CrossRef]
2. He, Y.; Bai, X.L.; Xiao, Q.L.; Liu, F.; Zhou, L.; Zhang, C. Detection of adulteration in food based on nondestructive analysis techniques: A review. *Crit. Rev. Food Sci. Nutr.* **2020**, *61*, 2351–2371. [CrossRef] [PubMed]
3. von Wuthenau, K.; Muller, M.-S.; Lina, C.; Marie, O.; Markus, F. Food authentication of almonds (*Prunus dulcis* Mill.). Fast origin analysis with laser ablation inductively coupled plasma mass spectrometry and chemometrics. *J. Agric. Food Chem.* **2022**, *70*, 5237–5244. [CrossRef]
4. Li, D.; Zang, M.W.; Wang, S.W.; Zhang, K.H.; Zhang, Z.Q.; Li, X.M.; Li, J.C.; Guo, W.P. Food fraud of rejected imported foods in China in 2009–2019. *Food Control* **2022**, *133*, 108619. [CrossRef]
5. Lim, K.; Pan, K.; Yu, Z.; Xiao, R.H. Pattern recognition based on machine learning identifies oil adulteration and edible oil mixtures. *Nat. Commun.* **2020**, *11*, 5353. [CrossRef]
6. Negi, A.; Pare, A.; Meenatchi, R. Emerging techniques for adulterant authentication in spices and spice products. *Food Control* **2021**, *127*, 108113. [CrossRef]
7. Czerwenka, C.; Muller, L.; Lindner, W. Detection of the adulteration of water buffalo milk and mozzarella with cow's milk by liquid chromatography-mass spectrometry analysis of beta-lactoglobulin variants. *Food Chem.* **2010**, *122*, 901–908. [CrossRef]
8. Acosta, G.; Arce, S.; Martinez, L.D.; Llabot, J.; Gomez, M.R. Monitoring of phenolic compounds for the quality control of melissa officinalis products by capillary electrophoresis. *Phytochem. Anal.* **2012**, *23*, 177–183. [CrossRef] [PubMed]
9. Hong, E.; Lee, S.Y.; Jeong, J.Y.; Park, J.M.; Kim, B.H.; Kwon, K.; Chun, H.S. Modern analytical methods for the detection of food fraud and adulteration by food category. *J. Sci. Food Agric.* **2017**, *97*, 3877–3896. [CrossRef]
10. Bian, X.H.; Lu, Z.K.; van Kollenburg, G. Ultraviolet-visible diffuse reflectance spectroscopy combined with chemometrics for rapid discrimination of Angelicae Sinensis Radix from its four similar herbs. *Anal. Methods* **2020**, *12*, 3499–3507. [CrossRef]

11. Torrecilla, J.S.; Rojo, E.; Dominguez, J.C.; Rodriguez, F. A novel method to quantify the adulteration of extra virgin olive oil with low-grade olive oils by UV-Vis. *J. Agric. Food Chem.* **2010**, *58*, 1679–1684. [CrossRef] [PubMed]
12. Li, X.; Zhang, L.X.; Zhang, Y.; Wang, D.; Wang, X.F.; Yu, L.; Zhang, W.; Li, P.W. Review of NIR spectroscopy methods for nondestructive quality analysis of oilseeds and edible oils. *Trends Food Sci. Technol.* **2020**, *101*, 172–181. [CrossRef]
13. Lohumi, S.; Lee, S.; Lee, H.; Cho, B.K. A review of vibrational spectroscopic techniques for the detection of food authenticity and adulteration. *Trends Food Sci. Technol.* **2015**, *46*, 85–98. [CrossRef]
14. Rivera-Perez, A.; Romero-Gonzalez, R.; Frenich, A.G. Feasibility of applying untargeted metabolomics with GC-Orbitrap-HRMS and chemometrics for authentication of black pepper (*Piper nigrum* L.) and identification of geographical and processing markers. *J. Agric. Food Chem.* **2021**, *69*, 5547–5558. [CrossRef] [PubMed]
15. Rios-Reina, R.; Garcia-Gonzalez, D.L.; Callejon, R.M.; Amigo, J.M. NIR spectroscopy and chemometrics for the typification of Spanish wine vinegars with a protected designation of origin. *Food Control* **2018**, *89*, 108–116. [CrossRef]
16. Chen, Z.W.; Harrington, P.D. Self-optimizing support vector elastic net. *Anal. Chem.* **2020**, *92*, 15306–15316. [CrossRef]
17. Jia, W.; Dong, X.Y.; Shi, L.; Chu, X.G. Discrimination of milk from different animal species by a foodomics approach based on high-resolution mass spectrometry. *J. Agric. Food Chem.* **2020**, *68*, 6638–6645. [CrossRef]
18. Shao, X.G.; Bian, X.H.; Liu, J.J.; Zhang, M.; Cai, W.S. Multivariate calibration methods in near infrared spectroscopic analysis. *Anal. Methods* **2010**, *2*, 1662–1666. [CrossRef]
19. Olivieri, A.C. Analytical advantages of multivariate data processing. One, two, three, infinity? *Anal. Chem.* **2008**, *80*, 5713–5720. [CrossRef]
20. Ying, Y.W.; Jin, W.; Yu, H.X.; Yu, B.W.; Shan, J.; Lv, S.W.; Zhu, D.; Jin, Q.H.; Mu, Y. Development of particle swarm optimization-support vector regression (PSO-SVR) coupled with microwave plasma torch-atomic emission spectrometry for quality control of ginsengs. *J. Chemom.* **2017**, *31*, e2862. [CrossRef]
21. Bian, X.H.; Diwu, P.Y.; Liu, Y.R.; Liu, P.; Li, Q.; Tan, X.Y. Ensemble calibration for the spectral quantitative analysis of complex samples. *J. Chemom.* **2018**, *32*, e2940. [CrossRef]
22. Li, Y.K.; Shao, X.G.; Cai, W.S. A consensus least squares support vector regression (LS-SVR) for analysis of near-infrared spectra of plant samples. *Talanta* **2007**, *72*, 217–222. [CrossRef] [PubMed]
23. Xu, L.; Ye, Z.-H.; Yan, S.-M.; Shi, P.-T.; Cui, H.-F.; Fu, X.-S.; Yu, X.-P. Combining local wavelength information and ensemble learning to enhance the specificity of class modeling techniques: Identification of food geographical origins and adulteration. *Anal. Chim. Acta* **2012**, *754*, 31–38. [CrossRef]
24. Hu, Y.; Peng, S.L.; Peng, J.T.; Wei, J.P. An improved ensemble partial least squares for analysis of near-infrared spectra. *Talanta* **2012**, *94*, 301–307. [CrossRef] [PubMed]
25. Granato, D.; Santos, J.S.; Escher, G.B.; Ferreira, B.L.; Maggio, R.M. Use of principal component analysis (PCA) and hierarchical cluster analysis (HCA) for multivariate association between bioactive compounds and functional properties in foods: A critical perspective. *Trends Food Sci. Tech.* **2018**, *72*, 83–90. [CrossRef]
26. Jiang, Y.Y.; Ge, H.Y.; Zhang, Y. Quantitative analysis of wheat maltose by combined terahertz spectroscopy and imaging based on Boosting ensemble learning. *Food Chem.* **2020**, *307*, 125533.
27. Bian, X.H.; Li, S.J.; Lin, L.G.; Tan, X.Y.; Fan, Q.J.; Li, M. High and low frequency unfolded partial least squares regression based on empirical mode decomposition for quantitative analysis of fuel oil samples. *Anal. Chim. Acta* **2016**, *925*, 16–22. [CrossRef]
28. Cadet, F.; Fontaine, N.; Vetrivel, I.; Chong, M.N.F.; Savriama, O.; Cadet, X.; Charton, P. Application of fourier transform and proteochemometrics principles to protein engineering. *BMC Bioinform.* **2018**, *19*, 382. [CrossRef]
29. Liu, Z.C.; Cai, W.S.; Shao, X.G. A weighted multiscale regression for multivariate calibration of near infrared spectra. *Analyst* **2009**, *134*, 261–266. [CrossRef]
30. Wu, X.Y.; Bian, X.H.; Lin, E.; Wang, H.T.; Guo, Y.G.; Tan, X.Y. Weighted multiscale support vector regression for fast quantification of vegetable oils in edible blend oil by ultraviolet-visible spectroscopy. *Food Chem.* **2021**, *342*, 128245. [CrossRef]
31. Perez-Canales, D.; Alvarez-Ramirez, J.; Jauregui-Correa, J.C.; Vela-Martinez, L.; Herrera-Ruiz, G. Identification of dynamic instabilities in machining process using the approximate entropy method. *Int. J. Mach. Tools Manu.* **2011**, *51*, 556–564. [CrossRef]
32. Liu, Y.; Cai, W.S.; Shao, X.G. Intelligent background correction using an adaptive lifting wavelet. *Chemom. Intell. Lab. Syst.* **2013**, *125*, 11–17. [CrossRef]
33. Wu, W.H.; Chen, C.C.; Jhou, J.W.; Lai, G. A rapidly convergent empirical mode decomposition method for analyzing the environmental temperature effects on stay cable force. *Comput. Aided. Civ. Infrastruct. Eng.* **2018**, *33*, 672–690. [CrossRef]
34. Dragomiretskiy, K.; Zosso, D. Variational mode decomposition. *IEEE. Trans. Signal Proces.* **2014**, *62*, 531–544. [CrossRef]
35. Chen, Q.M.; Chen, J.H.; Lang, X.; Xie, L.; Rehman, N.U.; Su, H.Y. Self-tuning variational mode decomposition. *J. Franklin Inst.* **2021**, *358*, 7825–7862. [CrossRef]
36. Lahmiri, S. Intraday stock price forecasting based on variational mode decomposition. *J. Comput. Sci. Neth.* **2016**, *12*, 23–27. [CrossRef]
37. Hu, H.L.; Wang, L.; Tao, R. Wind speed forecasting based on variational mode decomposition and improved echo state network. *Renew. Energ.* **2021**, *164*, 729–751. [CrossRef]
38. Wang, Z.J.; Wang, J.J.; Du, W.H. Research on fault diagnosis of gearbox with improved variational mode decomposition. *Sensors* **2018**, *18*, 3510. [CrossRef]

39. Thissen, U.; Pepers, M.; Ustun, B.; Melssen, W.J.; Buydens, L.M.C. Comparing support vector machines to PLS for spectral regression applications. *Chemom. Intell. Lab. Syst.* **2004**, *73*, 169–179. [CrossRef]
40. Nazari, M.; Sakhaei, S.M. Successive variational mode decomposition. *Signal Process.* **2020**, *174*, 107610. [CrossRef]
41. Tan, C.; Wang, J.Y.; Wu, T.; Qin, X.; Li, M.L. Determination of nicotine in tobacco samples by near-infrared spectroscopy and boosting partial least squares. *Vib. Spectrosc.* **2010**, *54*, 35–41. [CrossRef]

biosensors

MDPI

Article

Technology for Rapid Detection of Cyromazine Residues in Fruits and Vegetables: Molecularly Imprinted Electrochemical Sensors

Sihua Peng [1,2,†], Aqiang Wang [1,†], Yuyang Lian [1,†], Jingjing Jia [3,†], Xuncong Ji [3], Heming Yang [1], Jinlei Li [1], Shuyan Yang [1], Jianjun Liao [1,*] and Shihao Zhou [1,*]

1 Key Laboratory of Germplasm Resources Biology of Tropical Special Ornamental Plants of Hainan Province, College of Forestry, Hainan University, Haikou 570228, China; pengsihua@126.com (S.P.); fly88200939@163.com (A.W.); l18876114522@126.com (Y.L.); yangheming97@163.com (H.Y.); jlli1013@163.com (J.L.); ysy000567@foxmail.com (S.Y.)
2 College of Plant Protection, Hainan University, Haikou 570228, China
3 Institute of Plant Protection, Hainan Academy of Agricultural Sciences (Research Center of Quality Safety and Standards for Agricultural Products of Hainan Academy of Agricultural Sciences), Haikou 571199, China; j9405136318@163.com (J.J.); insects99@163.com (X.J.)
* Correspondence: liaojianjun008@163.com (J.L.); zsh88200939@126.com (S.Z.)
† These authors contributed equally to this work.

Abstract: Cyromazine is an insect growth regulator insecticide with high selectivity and is widely used in the production and cultivation of fruits and vegetables. In recent years, incidents of excessive cyromazine residues in food have occurred frequently, and it is urgent to establish an accurate, fast, and convenient method for the detection of cyromazine residues to ensure the safety of edible agricultural products. To achieve rapid detection of cyromazine residues, we prepared a molecularly imprinted electrochemical sensor for the detection of cyromazine residues in agricultural products. Samples of tomato (*Lycopersicon esculentum* Miller), cowpea (*Vigna unguiculata*), and water were tested for the recovery rate of cyromazine. The results showed that the concentration of cyromazine showed a good linear relationship with the peak response current of the sensor developed in this study. The lower limit of detection for cyromazine was 0.5 μmol/L, and the sensor also had good reproducibility and interference resistance. This paper can be used as a basis for the study of methods for the detection of cyromazine residues in edible agricultural products.

Keywords: edible agricultural products; rapid detection of pesticide residues; molecular imprinting sensor; cyromazine

Citation: Peng, S.; Wang, A.; Lian, Y.; Jia, J.; Ji, X.; Yang, H.; Li, J.; Yang, S.; Liao, J.; Zhou, S. Technology for Rapid Detection of Cyromazine Residues in Fruits and Vegetables: Molecularly Imprinted Electrochemical Sensors. *Biosensors* **2022**, *12*, 414. https://doi.org/10.3390/bios12060414

Received: 16 May 2022
Accepted: 9 June 2022
Published: 14 June 2022

1. Introduction

Cyromazine is an insect growth regulator insecticide with strong systemic, stomach poisoning, and contact killing abilities. It is currently widely used in the control of fly pests in fruits and vegetables such as cowpea (*Vigna unguiculata*) and tomato (*Lycopersicon esculentum* Miller) [1,2]. Cyromazine itself has low toxicity to humans, but its degradation product melamine, when ingested in large quantities, can cause irreversible damage to the kidneys and is a serious health hazard [3]. As modern technology and analytical methods continue to improve, there is a need for greater accuracy and sensitivity in the detection of cyromazine in food [4]. To prevent cyromazine residues in food from causing harm to consumers, it is necessary to develop a simple and rapid detection method for cyromazine residues with high sensitivity and accuracy.

At present, the relatively mature technologies for the detection of cyromazine residues in food include liquid chromatography–tandem mass spectrometry (LC-MS), ultra-high-performance liquid chromatography (UPLC), enzyme-linked immunosorbent assay (ELISA), and high-performance liquid chromatography (HPLC) [5–7]. Although the above-mentioned methods have high accuracy and sensitivity, the high equipment cost and complicated operation hinder their further development. Therefore, there is an urgent need to develop an analytical method for pesticide residue determination with good sensitivity and selectivity, low cost, easy portability, and large-scale use.

Molecularly imprinted polymers (MIPs) are an advanced means of overcoming the limitations of traditional detection methods described above. MIPs are artificial recognition materials that are complementary in size and shape to template molecules and can be used for the specific target purpose of molecule identification [8,9]. As a novel material and bionic molecular recognition element, MIPs have unique advantages over traditional enzymes, antibodies, and receptors, such as excellent stability, high selectivity, ease of preparation, stability in organic solvents, and broad compatibility with many scientific fields [10–12]. In addition to their excellent identification properties, MIPs offer the advantages of low cost, heat and pressure resistance, storage stability, and suitability in harsh chemical media. Compared to conventional detection techniques, electrochemical detection methods are simple, low cost, and have a fast response time, thus offering significant advantages for pesticide residue detection in food [13–15]. Pan et al. used coumarin as a template molecule and MIP as a recognition element to prepare a chemiluminescence sensor on a 96-well microplate for the determination of organophosphorus residues in milk samples. The synthesized MIP can specifically recognize seven organic phosphorus [16]. The lowest detection limit of the sensor was 1 pg/mL, and the recoveries for seven organophosphorus species ranged from 86.1% to 86.5%. This shows that the detection of pesticide residues in food by molecularly imprinted polymers has broad application prospects [17,18]. There is no report on the use of molecularly imprinted sensors for the detection of cyromazine residues in food. In this study, electrochemical detection of cyromazine was achieved for the first time, which provides a reference for the rapid detection of triazine pesticides and other pesticides in food. In this study, a portable molecular imprinting sensor for cyromazine was prepared by electrochemistry combined with a molecular imprinting technique, using a gold nanoparticle-modified electrode, cyromazine as a template molecule, α-methacrylic acid as a functional monomer, trimethylolpropane trimethacrylate as a crosslinking agent, and HCl as an eluent under specific conditions, and the method was applied to the rapid analysis and detection of cyromazine in tomato, cowpea, and water samples.

2. Materials and Methods

2.1. Materials and Reagents

Cyromazine was purchased from Zhengzhou Labor Agrochemicals Co., Ltd., Zhengzhou, China); concentrated hydrochloric acid, sodium dihydrogen phosphate, dibasic sodium phosphate, glutaric dialdehyde, and potassium chloride were purchased from Sinopharm Chemical Reagent Co., Ltd. (Shanghai, China); potassium hexacyanoferrate(II) was purchased from Xilong Scientific Co., Ltd. (Guangzhou, China); potassium ferricyanide was purchased from Guangzhou Chemical Reagent Factory (Guangzhou, China); α-methacrylic acid was purchased from Meryer (Shanghai, China) Chemical Technology Co., Ltd. (Shanghai, China); trimethylolpropane trimethacrylate was purchased from Hubei Shishun Bio-Technology Co., Ltd. (Huanggang, China); acetonitrile was purchased from Shanghai Macklin Biochemical Technology Co., Ltd. (Shanghai, China). All the above reagents were analytically pure. Cowpea and tomato were purchased at a nearby farmers' market. Chloroauric acid (99%) was purchased from Hubei Guangao Biotechnology Co., Ltd. (Wuhan, China). Water samples were collected from paddy fields (19.507036° N, 109.504222° E), rivers (19.513888° N, 109.4917.2° E), and botanical gardens (19.5124° N, 109.4988° E).

2.2. Instrumentations and Equipment

Electrochemical workstation (CHI660E, CH Instruments Ins., Austin, TX, USA), screen-printed electrodes (Qingdao Botan Technology Co., Ltd., Qingdao, China), 1/10,000 analytical balance (Quintix124-1CN, Sartorius, Germany), Neofuge 23R high-speed refrigerated centrifuge (Likang Biomedical Technology Holdings Co., Ltd., Hong Kong, China), FST-111-TH100 ultra-pure water machine (Thermo, Waltham, MA, USA).

2.3. Preparation of Solutions and Samples

2.3.1. Preparation of Pesticide Standard Solutions

First, 1.2374 g of cyromazine was placed in a 1000 mL volumetric flask, and a 6 μmol/L pesticide standard solution with phosphate-buffered saline (PBS) (pH = 7.4) solvent was prepared and stored in a room-temperature environment. The 6 μmol/L pesticide stock solutions were then diluted with PBS into a series of pesticide standard solutions of 0.5, 1.0, 1.5, 2.0, 2.5, and 3.0 μmol/L, respectively.

2.3.2. Pretreatment of Test Samples

Tomato samples: Prepared tomatoes were washed and dried, chopped, and beaten well. Thirty grams of the homogenate was weighed and added to 30 mL of acetonitrile solution and vortexed for 3 min, then poured into a 50 mL centrifuge tube and centrifuged at 8000 r/min for 5 min, and the supernatant was filtered for use. The corresponding amount of cyromazine was added to the supernatant to prepare sample solutions containing 1, 2, and 3 μmol/L cyromazine. Cowpea samples were treated the same as tomatoes.

Water samples: The sampling bottles were first rinsed with water and filled carefully so that they overflowed to avoid trapping air bubbles in the sealed bottle. The samples were shipped in a refrigerated box with an ice pack. Container preparation included washing with detergent, rinsing with tap and ultra-pure water, and finally air drying. After the samples were transported to the laboratory, they were stored at 4 °C, and the collected water samples were simply filtered within 48 h, and then the corresponding amount of cyromazine was added to prepare sample solutions containing 1, 2, and 3 μmol/L cyromazine.

2.4. Preparation of the Modifier

$HAuCl_4$ stock solution: 0.5 mol/L H_2SO_4 was used as a solvent and a quantitative amount of $HAuCl_4$ was added to configure a deposition solution containing 0.2% $HAuCl_4$.

Cyromazine and α-methacrylic acid stock solution: Cyromazine and α-methacrylic acid stock solution was prepared with PBS solvent in a molar ratio of 1:4. Then, 0.4 μmol α-methacrylic acid was added to a 100 mL volumetric flask and mixed with ultrasound for 30 min, 0.1 μmol of cyromazine stock solution was added, and then PBS solution was added to increase the volume to 100 mL, sonicated for 1 h, and stored at a temperature of 2 °C.

2.5. Preparation of Molecularly Imprinted Sensors

Referring to the methods of preparing electrodes by Shi et al. and Li et al. [19,20] and optimizing them, 80 μL of 0.2% $HAuCl_4$ was dropped on the working surface of the electrode, and deposited by the potentiostat method (voltage: −0.25 V) for 3 min. The surface was rinsed with water and dried. Then, 80 μL of the polymerization solution of cyromazine and α-methacrylic acid was added to the surface of the dried electrode and deposited at a voltage of −1.0 V for 5 min. At this time, the surface of the electrode was simultaneously polymerized with cyromazine and α-methacrylic acid, and the electrode was dried. After that, 10 μL of 2 mg/L trimethylolpropane trimethacrylate solution was added to crosslink cyromazine and α-methacrylic acid on the electrode surface. After drying for 12 h, 80 μL of 1% HCL solution was added to elute for 15 min under the condition of a potential range of −0.4 to +0.8 V to obtain a molecularly imprinted sensor. The sensor prepared with the polymer solution without cyromazine was used as the non-molecularly imprinted sensor (NIP). Figure 1 is a flow chart of the development and detection of the sensor.

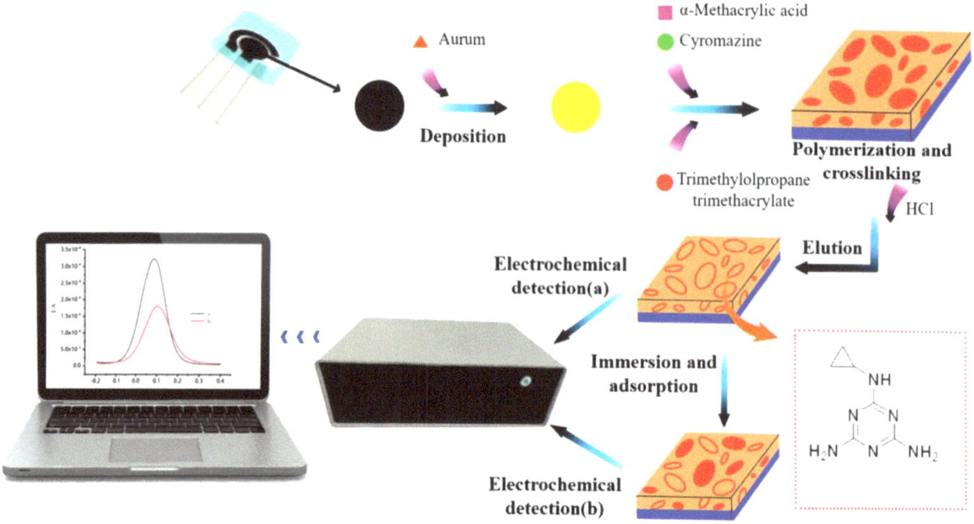

Figure 1. Flow chart of sensor development and testing. (**a**) The prepared molecularly imprinted sensor. (**b**) Molecularly imprinted sensor soaked in the sample solution.

2.6. *Electrochemical Characterization and Performance Testing of Sensors*

2.6.1. CV and EIS Characterization of Sensors

The homemade sensor was immersed in a 0.1 mol/L KCl solution containing 5.0 mmol/L [$K_3Fe(CN_6)$] and subjected to cyclic voltammetry (CV) scan for 2 segs at a potential range of −0.4 to +0.6 V to obtain the cyclic voltammogram of the sensor; the electrochemical impedance spectrum (EIS) of the sensor was obtained using the AC impedance method between 10^{-1} and 10^{-5} Hz.

2.6.2. Scanning Electron Microscope Characterization of Sensors

The homemade sensor was scanned under a scanning electron microscope to obtain a scanning electron microscope image of the sensor, and its surface was observed and analyzed.

2.6.3. DPV Performance Testing of Sensors

The prepared sensor was tested using differential pulse voltammetry (DPV) using 5.0 mmol/L [$K_3Fe(CN_6)$] and 0.1 mol/L KCL solution as the electrolyte solution to record the peak current (I_0) at this time [21]. Subsequently, the electrodes were immersed in a solution containing different concentrations of cyromazine from high to low, soaked for 20 min, and then taken out to dry, and the electrode peak current was tested with differential pulse voltammetry (DPV) at this time, which is recorded as I. The relative suppression ($I\%$) of different concentrations of cyromazine on the sensor was calculated using Equation (1):

$$I\% = \frac{I_0 - I}{I_0} \times 100\%. \quad (1)$$

Note: I_0: differential pulse voltammetry peak current of the sensor without pesticide immersion; I: differential pulse voltammetry peak current of the sensor with different concentrations of pesticide immersion; $I\%$: relative suppression of this sensor with different concentrations of cyromazine.

2.6.4. Repeatability Testing

The prepared sensor was soaked in PBS buffer solution containing 2 μmol/L of cyromazine, taken out and dried after 20 min, and 60 μL of 0.1 mol/L KCL solution containing 5.0 mmol/L [$K_3Fe(CN_6)$] was added to the working area of the sensor, and the differential pulse voltammetry (DPV) method was used to record the peak current value. The above operation was repeated six times and the relative standard deviation (RSD) was calculated.

2.6.5. Interference Test

Two pesticides, atrazine and metolachlor, which have a similar structure to cyromazine, were selected as the pesticides for resistance to interference. To 1 μmol/L of cyromazine solution, atrazine and metolachlor were added to form a mixture of 5 μmol/L, 10 μmol/L, 20 μmol/L, and 30 μmol/L of cyromazine solution, respectively, and this mixture was used as the interference solution. The sensor was immersed in the above solutions in a gradient order for 20 min, removed, and dried, then 60 μL of 0.1 mol/L KCl solution containing 5.0 mmol/L [$K_3Fe(CN_6)$] was added dropwise and scanned by differential pulse voltammetry, and the peak current values were recorded.

2.6.6. Actual Sample Recovery Testing

The sensor was immersed in the sample solution to be tested at different concentrations of cyromazine, left to adsorb for 20 min, and then removed and dried, then 60 μL of 0.1 mol/L KCl solution containing 5.0 mmol/L [$K_3Fe(CN_6)$] was dropped and scanned by differential pulse voltammetry, the peak current values were recorded, three replicates were made for each sample at each concentration, and the recoveries and relative standard deviations were calculated.

3. Results and Analysis

3.1. Analysis of the Results of Electrochemical Characterization

Figure 2A shows the result of the cyclic voltammetry in the sensor preparation steps where, after the screen-printed electrode is modified with chloroauric acid, the peak current of the electrode increases significantly, indicating that the gold nanoparticles are successfully deposited on the surface of the electrode, thereby increasing the conductivity of the electrode. When the polymer of cyromazine and α-methacrylic acid is deposited on the surface of the electrode, the presence of the polymer hinders the diffusional permeability of the redox marker, resulting in a decrease in the peak current of the electrode. After being eluted by HCL, some of the cyromazine molecules in the polymer were eluted, the gap between the polymer molecules became larger, and the electron transport was accelerated. Non-molecularly imprinted electrochemical sensors cannot crosslink to form long chains due to the absence of the participation of cyromazine in the preparation process, resulting in the inability of the polymerization solution to crosslink to form a long chain, which makes the gap between molecules smaller, resulting in a weakening of the conductivity of the sensor. After HCL elution, the peak current of the molecularly imprinted electrochemical sensor is greater than that of the non-molecularly imprinted electrochemical sensors, indicating that the blotting site on the surface of the sensor has good recognition performance for cyromazine.

The electrochemical impedance spectroscopy was performed on the above electrodes respectively, and the results are shown in Figure 2B. The results showed that the electron transport was promoted and the impedance spectrum radius was significantly reduced after chloroauric acid deposition. After the polymer was deposited and eluted, the impedance spectrum radius increased, indicating that the cyromazine and the α-methacrylic acid polymer were successfully bound to the sensor.

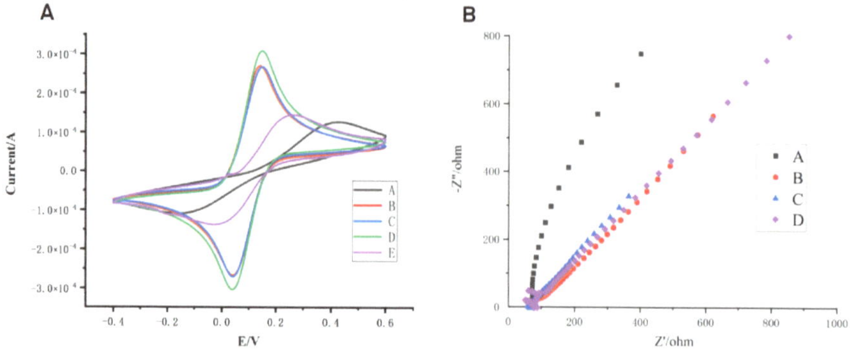

Figure 2. Structural characterization of the sensor in 5.0 mmol/L [$K_3Fe(CN_6)$] solution in 0.1 mol/L KCl. (**A**) Cyclic voltammogram (CV): A bare electrode CV (bare SPCE), B electrode after deposition of $HAuCl_4$ solution, C electrode after polymerizing cyromazine molecular polymer, D electrode after elution, E electrode (CK) after elution of non-imprinted sensor. (**B**) Electrochemical impedance spectroscopy (EIS): A bare electrode, B electrode after deposition of $HAuCl_4$ solution, C electrode after polymerizing cyromazine molecular polymer, D electrode after elution.

3.2. Characterization Results of the Sensors Using Scanning Electron Microscopy

The surface of the electrode modified with chloroauric acid had many small particles (Figure 3B) and became smooth, indicating that the gold particles were successfully assembled on the sensor. After elution with 1% HCl, the electrode surface became flat and smooth, and the surface lumps were reduced (Figure 3D), indicating that the cyromazine molecules were successfully eluted.

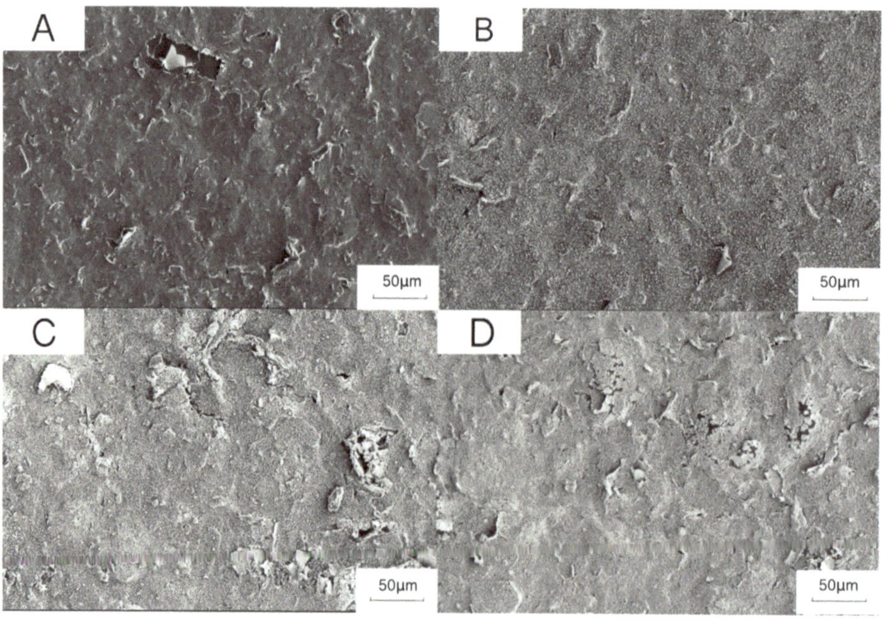

Figure 3. Scanning electron micrograph of the sensor. (**A**) Bare electrode CV (bare SPCE), (**B**) electrode after deposition of $HAuCl_4$ solution, (**C**) electrode after polymerizing cyromazine molecular polymer, (**D**) electrode after elution.

3.3. Performance Test Results of Sensors

The prepared sensor was used to test solutions containing different concentrations of cyromazine, the results are shown in Figure 4A, and the relative suppression curve of the cyromazine solution was drawn in combination with the peak current of the sensor (Figure 4B). It can be seen that the peak current of the sensor is negatively correlated with the concentration of cyromazine, and the specific performance is that with the increase in the concentration of cyromazine, the corresponding peak current decreases. Therefore, in a certain concentration range, there is a good linear relationship between the response current of the sensor and the pesticide concentration.

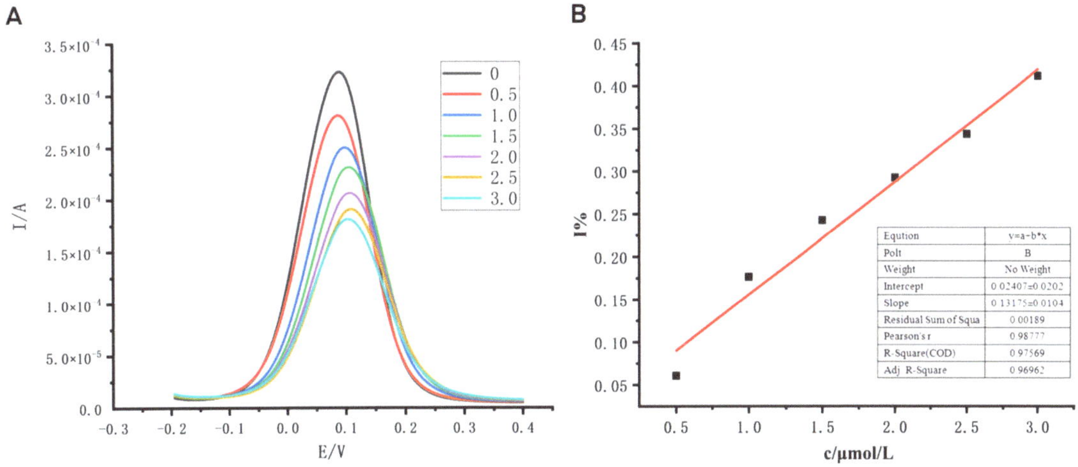

Figure 4. Standard curve and relative suppression curve of cyromazine solution with different concentrations. Note: (**A**) DPV curve of the molecularly imprinted sensor with different concentrations of cyromazine, the (**B**) curve of relative suppression of the molecularly imprinted sensor with different concentrations of cyromazine; 0, 0.5, 1.0, 1.5, 2.0, 2.5, 3.0 are different concentrations of cyromazine, where the concentration unit is μmol/L.

3.4. Results of Repeatability Tests

We found that the current response signal of the sensor did not show significant attenuation during the test, and the relative standard deviation of the results of the first six tests was 4.56% (Figure 5), indicating that the results obtained by the prepared sensor in the six consecutive tests were more accurate.

3.5. Anti-Interference Test Results

Atrazine and metolachlor were added to 1 μmol/L cyromazine solution to prepare cyromazine containing 5 μmol/L, 10 μmol/L, 20 μmol/L, 30 μmol/L atrazine and metolachlor amine solution, respectively, and each group of solutions was tested. The results show that adding different concentrations of interfering substances has little effect on the test results, and the difference between the relative suppression and the original solution is less than 5% (Tables 1 and 2). The above results show that this molecularly imprinted sensor has good anti-interference performance.

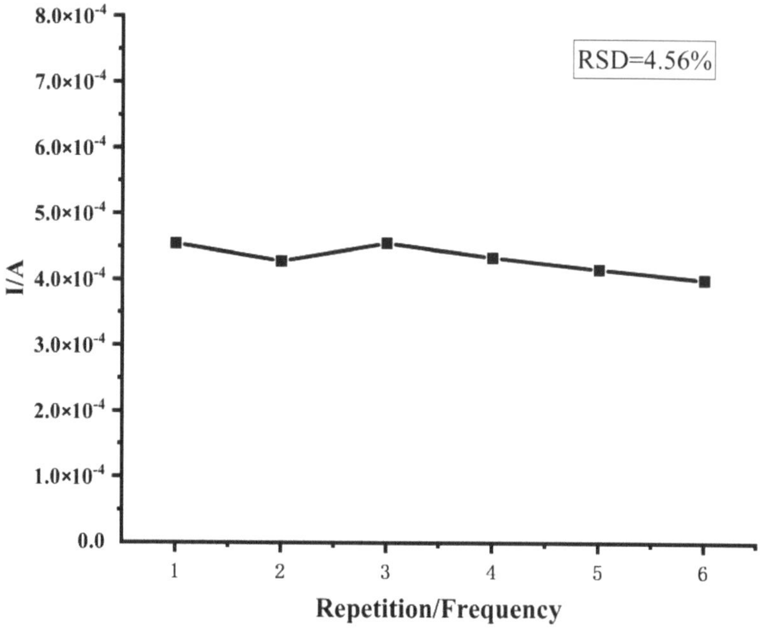

Figure 5. Results of the sensor repeatability test.

Table 1. Test results of the sensor's resistance to atrazine interference.

Samples	1 μmol/L Cyromazine Solution	1 μmol/L Cyromazine Solution + 5 μmol/L Atrazine	1 μmol/L Cyromazine Solution + 10 μmol/L Atrazine	1 μmol/L Cyromazine Solution + 20 μmol/L Atrazine	1 μmol/L Cyromazine Solution + 30 μmol/L Atrazine
Relative suppression	11.7%	9.07%	11.73%	8.64%	13.57%
Difference between relative suppression and stock solution	0.00%	2.63%	0.03%	3.06%	1.87%

Table 2. The test results of the sensor's resistance to metolachlor interference.

Samples	1 μmol/ Cyromazine Solution	1 μmol/L Cyromazine Solution + 5 μmol/L Metolachlor	1 μmol/L Cyromazine Solution + 10 μmol/L Metolachlor	1 μmol/L Cyromazine Solution + 20 μmol/L Metolachlor	1 μmol/L Cyromazine Solution + 30 μmol/L Metolachlor
Relative suppression	17.40%	17.20%	19.07%	15.36%	16.27%
Difference between relative suppression and stock solution	0%	0.20%	1.67%	2.04%	0.93%

3.6. Recovery Analysis of Actual Samples

The prepared sensors were used to detect the recovery of tomato and cowpea and water samples, and each sample was repeated three times for each concentration. The results are shown in Tables 3 and 4. The recovery rates of tomato and cowpea were 90.14% to 101.67% and 90.64% to 101.10%. The spiked recoveries of the water samples ranged from 91.1% to 108%, 114% to 118%, and 92.5% to 97.1%, respectively, and the relative standard deviations were all less than 6%. This shows that the molecularly imprinted sensor can meet the requirements of rapid detection of cyromazine in tomato and cowpea and water samples [22].

Table 3. Results of actual sample recovery.

Samples	Added (μmol/L)	Found (μmol/L)	Recovery (n = 3)	RSD (n = 3)
Tomato (*Lycopersicon esculentum*)	1	1	99.89%	4.16%
	2	2.03	101.67%	1.56%
	3	2.70	90.14	2.61%
Cowpea (*Vigna unguiculata*)	1	0.90	94.7%	4.98%
	2	2.02	101.10%	1.66%
	3	2.72	90.64%	2.52%

Table 4. Results of recovery of different water samples.

Samples	Added (μmol/L)	Found (μmol/L)	Recovery (n = 3)	RSD (n = 3)
River water	1	1.08	108%	2.39%
	2	1.82	91.1%	1.28%
	3	2.95	98.3%	3.39%
Water in the paddy field	1	1.16	116%	0.9%
	2	2.37	118%	2.33%
	3	3.43	114%	2.02%
Water in the botanical garden	1	0.925	92.5%	0.87%
	2	1.91	95.6%	5.79%
	3	2.92	97.4%	5.77%

4. Discussion

Pretreatment of samples with a large number of organic solvents leads to the presence of many interfering substances in the treated samples, which have an unavoidable effect on the determination [23,24]. In this study, we selected two pesticides structurally similar to mefenoxam for interference experiments and found that the peak currents of the sensors did not change significantly after the addition of the interfering substances to the solution compared to those without the addition of the interfering substances. This suggests that molecularly imprinted polymers have the property of specific recognition. Li et al. [25] prepared a biosensor with a dual recognition system to detect lincomycin in meat products and the environment, with a detection limit of 1.6×10^{-13} mol/L, which meets the requirement of lincomycin in daily meat products. Dinali et al. synthesized a mesoporous molecularly imprinted polymer (core@mMIP) on the surface of silica nanoparticles to use it as a filler sorbent for microextraction for the selective determination of pesticides in apple juice. The sensor has been successfully applied to real samples of processed and fresh apple juice [26]. Li et al. successfully synthesized novel core–shell structured zeolite imidazole skeleton-8@ molecularly imprinted polymers by a surface imprinting technique and used them as sorbents for solid-phase extraction of organophosphorus pesticides. Under optimal conditions, the detection range of the method was from 1 to 200 μg/L. The recoveries of three different concentrations spiked in agricultural products (cauliflower, radish, pear, melon cauliflower, radish, pear, and muskmelon) ranged from 82.5% to 123.0% with the relative standard deviations below 8.24% [27]. These show that molecularly imprinted polymers have excellent recognition ability, and based on this property, molecularly imprinted sensors will be more rapidly developed and applied in the field of rapid detection of pesticide residues in food [28–30].

At present, the traditional pesticide residue detection technology has formed a complete set of detection systems and is widely used to detect pesticide residues in food [31–33]. Yu et al. [33] used a fluorescent quantitative method to detect paclobutrazol pesticide residues in apples. At a wavelength of 341 nm, the average recovery rate of the samples was 99.62%, and the relative standard deviation was 0.52%. Tsochatzis et al. [34] used matrix solid-phase dispersive extraction and high-performance liquid chromatography to detect eight pesticides commonly used in rice. The detection limits ranged from 0.002 mg/L to 0.2 mg/L with a relative standard deviation of less than 12%, which met the conditions for the detection of pesticide residues in rice. Although these traditional laboratory assays are characterized by high sensitivity and accurate results, they require complex sample pretreatment, have a high cost of detection, and are inconvenient in terms of portability, which hinder their further development. In this study, we prepared molecularly imprinted sensors with spiked recoveries ranging from 90% to 102% with relative deviations (RSDs) <5% using cyromazine as the template molecule and α-methacrylic acid as the functional monomer. The detection limit of the sensor was 0.083 mg/L, meeting the demand for the detection of cyromazine residues in fruits and vegetables. When we tested the repeatability of the sensor, we found that there was no significant difference (RSD < 5%) between the results obtained when the same sensor was used six times for the same concentration of the solution. This result indicates that the same sensor can be used at least six times, thus not only reducing the cost of pesticide residue testing but also reducing the number of contaminants generated during the testing process. The experimental results show that compared with traditional detection methods, the molecularly imprinted electrochemical sensor not only has accurate results and high sensitivity, but also reduces detection costs, is portable, suitable for large-scale applications, and has broad market prospects [35,36].

In this study, the sensor was only used to detect cyromazine residues in cowpea, tomato, and water samples, so the detection range needs to be expanded. Since cyromazine is also commonly used in the production and cultivation of edible agricultural products such as celery (*Apium graveolens* L.), cucumber (*Cucumis sativus* L.), and mango (*Mangifera indica* L.), this sensor can also be used to detect cyromazine residues in other foods to ensure food quality and consumer safety.

Author Contributions: S.P., A.W., Y.L., J.J., J.L. (Jinlei Li) and S.Z. participated in the study design and analysis of the manuscript. X.J., H.Y., S.Y. and J.L. (Jianjun Liao) participated in the study design and helped to draft the manuscript. S.Z. revised and processed the manuscript and provided supervision and financial support. All authors have read and agreed to the published version of the manuscript.

Funding: This study was supported by Hainan Province Science and Technology Special Fund (ZDYF2022XDNY163), Hainan Province Science and Technology Special Fund (ZDYF2022SHFZ094), National Natural Science Foundation of China (22166016), High-level Talent Project of Hainan Natural Science Foundation (320RC499), Open project of Hainan Key Laboratory for Control of Plant Diseases and Insect Pests. The Project of Sanya Yazhou Bay Science and Tech-nology City.

Institutional Review Board Statement: This is an observational study. The Insects Research Ethics Committee has confirmed that no ethical approval is required.

Data Availability Statement: The datasets generated analyzed during the current study are available from the corresponding author on reasonable request.

Conflicts of Interest: The authors declare no conflict of interest.

References

1. Fu, D.; Zhang, Y.; Zhang, S.; Wang, M.; Liang, X.Y.; Zhang, Y.Q.; Huang, S.H.; Zhang, C.H. Validation and application of a QuEChERS- based method for estimation of the half-lives of cyromazine and acetamiprid in cowpeas and soil by LC-ESI-MS/MS. *Int. J. Environ. Anal. Chem.* **2022**, *102*, 650–666. [CrossRef]
2. Mansoor, M.M.; Shad, S.A. Risk assessment of cyromazine and methoxyfenozide resistance suggests higher additive genetic but lower environmental variation supporting quick resistance development in non-target *Chrysoperla carnea* (Stephens). *Environ. Monit. Assess.* **2022**, *194*, 66. [CrossRef] [PubMed]
3. Xing, H.B.; Zheng, B.W.; Li, X.Y.; Dang, X.; Zhang, H.B.; Tian, F.X.; Hu, X.J. Sensitive SERS detection of melamine and cyromazine in raw milk using aptamer-based in situ silver nanoparticles synthesis. *Results Phys.* **2022**, *4*, 100266. [CrossRef]
4. Ferrer, C.; Lozano, A.; Uclés, S.; Valverde, A.; Fernández-Alba, A.R. European Union Proficiency Tests for pesticide residues in fruit and vegetables from 2009 to 2016: Overview of the results and main achievements. *Food Control* **2017**, *82*, 101–113. [CrossRef]
5. Pathak, A.K.; Viphavakit, C. A review on all-optical fiber-based VOC sensors: Heading towards the development of promising technology. *Sens. Actuator A Phys.* **2022**, *338*, 113455. [CrossRef]
6. Cennamo, N.; D'Agostino, G.; Perri, C.; Arcadio, F.; Chiaretti, G.; Parisio, E.M.; Vettori, C.; Marzo, F.D.; Cennamo, R.; Porto, G.; et al. Proof of concept for a quick and highly sensitive on-site detection of SARS-CoV-2 by plasmonic optical fibers and molecularly imprinted polymers. *Sensors* **2021**, *21*, 1681. [CrossRef]

7. Mousazadeh, M.; Jahangiri-Manesh, A.; Nikkhah, M.; Abbasian, S.; Moshaii, A.; Masroor, M.J.; Norouzi, P. Detection of hexanal gas as a volatile organic compound cancer biomarker using a nanocomposite of gold nanoparticles and selective polymers. *J. Electroanal. Chem.* **2022**, *905*, 115962. [CrossRef]
8. Wang, B.; Okoth, O.K.; Yan, K.; Zhang, J. A highly selective electrochemical sensor for 4-chlorophenol determination based on molecularly imprinted polymer and PDDA-functionalized graphene. *Sens. Actuators B Chem.* **2016**, *236*, 294–303. [CrossRef]
9. Xiao, T.; Huang, J.; Wang, D.; Meng, T.; Yang, X.R. Au and Au-Based nanomaterials: Synthesis and recent progress in electrochemical sensor applications. *Talanta* **2020**, *206*, 120210. [CrossRef]
10. Tokonami, S.; Shiigi, H.; Nagaoka, T. Review: Micro- and nanosized molecularly imprinted polymers for high-throughput analytical applications. *Anal. Chim. Acta* **2009**, *641*, 7–13. [CrossRef]
11. Afkhami, A.; Ghaedi, H.; Madrakian, T.; Ahmadi, M.; Mahmood-Kashani, H. Fabrication of a new electrochemical sensor based on a new nano-molecularly imprinted polymer for highly selective and sensitive determination of tramadol in human urine samples. *Biosens. Bioelectron.* **2013**, *44*, 34–40. [CrossRef] [PubMed]
12. Ge, L.; Wang, S.; Yu, J.; Li, N.; Ge, S.; Yan, M. Molecularly Imprinted Polymer Grafted Porous Au-Paper Electrode for a Microfluidic Electro-Analytical Origami Device. *Adv. Func.* **2013**, *23*, 3115–3123. [CrossRef]
13. Guth, U.; Vonau, W.; Zosel, J. Recent developments in electrochemical sensor application and technology- a review. *Meas. Sci. Technol.* **2009**, *20*, 91–100. [CrossRef]
14. Tian, X.K.; Liu, L.; Li, Y.; Yang, C.; Zhou, Z.X.; Nie, Y.L.; Wang, Y.X. Nonenzymatic electrochemical sensor based on CuO-TiO_2 for sensitive and selective detection of methyl parathion pesticide in ground water. *Sens. Actuators B Chem.* **2018**, *256*, 135–142. [CrossRef]
15. Hu, H.; Yang, L. Development of enzymatic electrochemical biosensors for organophosphorus pesticide detection. *J. Environ. Sci. Health B* **2020**, *56*, 1–13. [CrossRef]
16. Pan, Y.C.; Liu, X.; Liu, J.; Wang, J.P.; Liu, J.X.; Gao, Y.X.; Ma, N. Chemiluminescence sensors based on molecularly imprinted polymers for the determination of organophosphorus in milk. *J. Dairy Sci.* **2022**, *105*, 3019–3031. [CrossRef]
17. Gholivand, M.B.; Torkashvand, M.; Malekzadeh, G. Fabrication of an electrochemical sensor based on computationally designed molecularly imprinted polymers for determination of cyanazine in food samples. *Anal. Chim. Acta* **2012**, *713*, 36–44. [CrossRef]
18. Kan, X.; Zhou, H.; Li, C.; Zhu, A.; Xing, Z.; Zhao, Z. Imprinted electrochemical sensor for dopamine recognition and determination based on a carbon nanotube/ polypyrrole film. *Electrochim. Acta* **2012**, *63*, 69–75. [CrossRef]
19. Shi, Z.H.; Lu, Y.L.; Chen, Z.T.; Cheng, C.; Xu, J.; Zhang, Q.Q.; Yan, Z.P.; Luo, Z.S.; Liu, Q.J. Electrochemical non-enzymatic sensing of glycoside toxins by boronic acid functionalized nano-composites on screen-printed electrode. *Sens. Actuators B Chem.* **2021**, *329*, 129197. [CrossRef]
20. Li, Y.G.; Chen, X.; Ren, H.L.; Li, X.; Chen, S.Y.; Ye, B.C. A novel electrochemical sensor based on molecularly imprinted polymer-modified C-ZIF67@Ni for highly sensitive and selective determination of carbendazim. *Talanta* **2022**, *237*, 122909. [CrossRef]
21. Peng, S.H.; Yang, S.Y.; Zhang, X.; Jia, J.J.; Chen, Q.L.; Lian, Y.Y.; Wang, A.Q.; Zeng, B.; Yang, H.M.; Li, J.L.; et al. Analysis of imidacloprid residues in mango, cowpea and water samples based on portable molecular imprinting sensors. *PLoS ONE* **2021**, *16*, e0257042. [CrossRef]
22. Calderon, R.; García-Hernández, J.; Palma, P.; Leyva-Morales, J.B.; Zambrano-Soria, M.; Bastidas-Bastidas, P.J.; Godoy, M. Assessment of pesticide residues in vegetables commonly consumed in Chile and Mexico: Potential impacts for public health. *J. Food Compost. Anal.* **2022**, *108*, 104420. [CrossRef]
23. Li, G.L.; Qi, X.M.; Wu, J.T.; Xu, L.J.; Wan, X.; Liu, Y.; Chen, Y.W.; Li, Q. Ultrasensitive, label-free voltammetric determination of norfloxacin based on molecularly imprinted polymers and Au nanoparticle-functionalized black phosphorus nanosheet nanocomposite. *J. Hazard. Mater.* **2022**, *436*, 129107. [CrossRef] [PubMed]
24. Li, G.L.; Qi, X.M.; Zhang, G.Q.; Wang, S.L.; Li, K.H.; Wu, J.T.; Wan, X.; Liu, Y.; Li, Q. Low-cost voltammetric sensors for robust determination of toxic Cd(II) and Pb(II) in environment and food based on shuttle-like α-Fe2O3 nanoparticles decorated β-Bi2O3 microspheres. *Microche. J.* **2022**, *179*, 107515. [CrossRef]
25. Li, S.H.; Liu, C.H.; Yin, G.H.; Zhang, Q.; Luo, J.H. Aptamer-molecularly imprinted sensor base on electrogenerated chemiluminescence energy transfer for detection of lincomycin. *Biosens. Bioelectro.* **2017**, *91*, 687–691. [CrossRef]
26. Dinali, L.A.F.; de Oliveira, H.L.; Teixeira, L.S.; Borges, W.S.; Borges, K.B. Mesoporous molecularly imprinted polymer core@shell hybrid silica nanoparticles as adsorbent in microextraction by packed sorbent for multiresidue determination of pesticides in apple juice. *Food Chem.* **2021**, *345*, 128745. [CrossRef]
27. Li, Q.; Wu, J.T.; Liu, Y.; Qi, X.M.; Jin, H.G.; Yang, C.; Liu, J.; Li, G.L.; He, Q.G. Recent advances in black phosphorus-based electrochemical sensors: A review. *Anal. Chim. Acta* **2021**, *1170*, 338480. [CrossRef]
28. Li, Y.; Li, B.Z.; Qi, Y.; Zhang, Z.P.; Cong, S.; She, Y.X.; Cao, X.L. Synthesis of metal-organic framework @molecularly imprinted polymer adsorbents for solid phase extraction of organophosphorus pesticides from agricultural products. *J. Chromatogr. B* **2022**, *1188*, 123081. [CrossRef]
29. Basak, S.; Venkatram, R.; Singhal, R.S. Recent advances in the application of molecularly imprinted polymers (MIPs) in food analysis. *Food Control* **2022**, *139*, 109074. [CrossRef]
30. Li, F.Z.; Ni, B.B.; Zheng, Y.R.; Huang, Y.X.; Li, G.L. A simple and efficient voltammetric sensor for dopamine determination based on ZnO nanorods/electro-reduced graphene oxide composite. *Surf. Interfaces* **2021**, *26*, 101375. [CrossRef]

31. Farooq, S.; Wu, H.Y.; Nie, J.Y.; Ahmad, S.; Muhammad, I.; Zeeshan, M.; Khan, R.; Asim, M. Application, advancement and green aspects of magnetic molecularly imprinted polymers in pesticide residue detection. *Sci. Total Environ.* **2022**, *804*, 150293. [CrossRef] [PubMed]
32. Kunpatee, K.; Kaewdorn, K.; Duangtong, J.; Chaiyo, S.; Chailapakul, O.; Kalcher, K.; Kerr, M.; Samphao, A. A new disp electrochemical sensor for the individual and simultaneous determination of carbamate pesticides using a nanocom modified screen-printed electrode. *Microchem. J.* **2022**, *177*, 107318. [CrossRef]
33. Yu, Y.; Qu, Y.; Zhang, M.; Guo, X.; Zhang, H. Fluorescence detection of paclobutrazol pesticide residues in apple juice. *Int.* **2020**, *224*, 165542. [CrossRef]
34. Tsochatzis, E.D.; Menkissoglu-Spiroudi, U.; Karpouzas, D.G.; Tzimou-Tsitouridou, R. A multi-residue method for pesti residue analysis in rice grains using matrix solid-phase dispersion extraction and high-performance liquid chromatography–dio array detection. *Anal. Bioanal. Chem.* **2010**, *397*, 2181–2190. [CrossRef] [PubMed]
35. Umapathi, R.; Ghoreishian, S.M.; Sonwal, S.; Rani, G.M.; Huh, Y.S. Portable electrochemical sensing methodologies for on-site detection of pesticide residues in fruits and vegetables. *Coord. Chem. Rev.* **2022**, *435*, 214305. [CrossRef]
36. Sohrabi, H.; Sani, P.S.; Orooji, Y.; Majidi, M.R.; Yoon, Y.; Khataee, A. MOF-based sensor platforms for rapid detection of pesticides to maintain food quality and safety. *Food Chem. Toxicol.* **2022**, *165*, 113176. [CrossRef]

MDPI AG
Grosspeteranlage 5
4052 Basel
Switzerland
Tel.: +41 61 683 77 34

Biosensors Editorial Office
E-mail: biosensors@mdpi.com
www.mdpi.com/journal/biosensors

www.ingramcontent.com/pod-product-compliance
Lightning Source LLC
Chambersburg PA
CBHW041342210926
51850CB00013B/2961
* 9 7 8 3 7 2 5 8 2 8 2 6 5 *